THE DOLPHIN

WHO SAVED ME

Photo of Jock. MIKE BOSSLEY

Melody Horrill

THE DOLPHIN WHO SAVED ME

How an Extraordinary Friendship

Helped Me Overcome

Trauma and Find Hope

GREYSTONE BOOKS
Vancouver/Berkeley/London

First published in Canada, the U.S. and the U.K.
by Greystone Books in 2023
Originally published in Australia as
*A Dolphin Called Jock: An Injured Dolphin, a Lost
Young Woman, a Story of Hope* by Allen & Unwin
Copyright © 2022 by Melody Horrill

23 24 25 26 27   5 4 3 2 1

Greystone Books Ltd.
greystonebooks.com

Cataloguing data available from Library and Archives Canada
ISBN 978-1-77840-052-0 (pbk.)
ISBN 978-1-77840-053-7 (epub)

Proofreading by Alison Strobel
Cover and interior design by Jessica Sullivan
Cover photo by Andrea Izzotti / Shutterstock
Photo on facing page by Mike Bossley

Printed and bound in Canada on FSC® certified paper
at Friesens. The FSC® label means that materials used
for the product have been responsibly sourced.

Greystone Books thanks the Canada Council for the Arts,
the British Columbia Arts Council, the Province of British
Columbia through the Book Publishing Tax Credit and the
Government of Canada for supporting our publishing activities.

MIX
Paper from
responsible sources
FSC® C016245
www.fsc.org

BRITISH COLUMBIA

BRITISH COLUMBIA
ARTS COUNCIL
An agency of the Province of British Columbia

Canada Council   Conseil des arts
for the Arts   du Canada

Greystone Books gratefully acknowledges the xʷməθkʷəy̓əm (Musqueam),
Sḵwx̱wú7mesh (Squamish), and səlilwətaɬ (Tsleil-Waututh) peoples on
whose land our Vancouver head office is located.

For the best teachers I ever had,
the Port River dolphins, especially Jock,
who taught me how to trust.

To the Dolphin alone, beyond all other, nature
has granted what the best philosophers seek:
friendship for no advantage.
Plutarch (46 CE–120 CE)

Jock leaping, rising out of the water
effortlessly and with great speed. NEWSPIX

CONTENTS

I hang on to the ladder of the research boat during one of my first swims with Jock, in one of the magical mangrove channels where we frequently interacted with him. MIKE BOSSLEY

PROLOGUE

IT WASN'T JUST the oppressive Adelaide heat that kept me tossing and turning all night. I was excited. In the morning, I was going out onto the Port River to visit dolphins with one of my university lecturers, Dr. Mike Bossley.

I'd never been before. The only real dolphins I'd seen were from on board a ship, travelling from Singapore to Australia as a child. I knew about them, however—I'd read stories of dolphins seeking human interaction, even forming special bonds with people. They seemed like inquisitive, intelligent creatures with a vibrancy and vitality that were irresistible.

Mike had been studying the Port River dolphins for years and knew many of the resident individuals by name. I couldn't wait to see them for myself. Even though it was voluntary, I was buzzing at the prospect of helping with the work. I wanted to learn more about these creatures.

WE ARE SETTING OFF in a small rubber dinghy under a smoggy, soft sienna sky flecked with cotton-white clouds. The briny air is mixed with a faint scent of decomposition from a nearby dump. Between the distant shrieks of gulls comes a barely audible mechanical hum.

My T-shirt is already damp. The air feels tacky, glutinous and clammy, which is unusual for South Australia. Apparently, tropical moisture has moved down into the state. A gossamer haze rises from the water, the groping tendrils of steam dissipating as they ascend. This part of the river is artificially warmed by the nearby power station, Mike had told me.

We putter away from the ramp. Crooked trees, leaves the colour of absinthe, cling to the riverbank with gangly spider-like legs, their convoluted roots jutting out of the mud, supporting contorted trunks with abundant canopy. The mangroves seem to be sitting in a sea of stalagmites, more suited to the subterranean. I later discover they are called pneumatophores, snorkel-like roots which protrude from the muddy riverbank and help the trees to breathe. They remind me of *The Day of the Triffids*, one of my favourite early sci-fi movies. These swamps are essential for the health of the ecosystem. They provide a rich and vital habitat for fish during their breeding cycle.

A movement catches my attention.

"Is that a dolphin?" I blurt, bouncing on the rubber seat, almost losing my balance.

Mike stops, stands and cuts the motor.

I glimpse what appears to be a dorsal fin, gliding past one of the boats anchored in the channel.

The dolphin keeps looping around the same boat. Around and around. Occasionally, he disappears underwater, only to reappear shortly after with a "puh" sound. He seems oblivious to our presence.

"Oh my God, what's wrong with his dorsal fin?" I ask. It doesn't look real. It seems as if it had been crudely fashioned by some wayward kid with playdough. It is contorted and twisted, and its tip appears ready to fall off at any moment.

"Yeah," Mike replies. "We know he's been tangled up in discarded nets and fishing lines a lot and it's cut into his flesh. He seems to be quite young, an adolescent, so he could've first been entangled as a baby and the line disfigured his fin as it was growing. Over the years the injuries have made it easy for him to get tangled up again. He spends most of his life just circling that boat."

Mike suggests I toss a paddle into the water. The dolphin's response to it amazes me. He seems suddenly exuberant and engrossed by this inanimate object. It makes me question even more why he is alone. Where are the other dolphins? I thought they lived in pods. I know they are highly social creatures that rely on one another. It seems odd to see one all alone, acting so strangely, so obsessively.

As he goes back to his ceaseless circling, this dolphin with the deformed fin looks so lonely and isolated. I feel compassion wash over me. He looks like an outcast. Like me, he seems damaged. I wonder if he too has internal scars, as well as external ones. He looks disconnected and alone, like me, wounded and adrift in the world. Maybe he'd been abandoned by his mother. Perhaps he'd suffered pain and been shunned by his peers, as I had. It was also possible he just didn't know how to interact with other dolphins, or was too afraid to try. Something deep inside feels an instant kinship with him. I sense the beginnings of a profound connection.

JOCK HAD THAT dolphin smile, that fixed expression making him always look happy. I knew what it was like to wear such a mask. Unlike him, however, I could choose to put mine on only when I needed to; it was a vital part of my armour. When I did, I felt no one could see beneath it. I didn't want anyone to breach the impenetrable plates I'd placed so carefully around my heart. Few people knew I carried a lifetime of sorrow. My family life had been dysfunctional, wracked by bouts of extreme violence and cruelty, ruled by fear.

I grew up doubting everything—myself, my family, other people, the world. I felt disconnected. The connections I did have were flimsy and superficial. I didn't understand love. I wasn't worthy. To me, love was just a concept. It was conditional, transactional, fraught with concealed traps and hidden clauses, something to be wary of. I knew something inside me was broken. I told myself to just live with it. I could conceal it, and maybe one day fix it. Problem was, I just didn't know how to do it.

Bobbing in concert with the current, I felt an overwhelming urge to reach out to this solitary creature circling the boat, to reassure him that he wasn't alone. I wanted to let him know somehow that I cared, that I understood. For the first time in my life, I felt compelled to nurture a relationship.

My response surprised me. The trip was meant to be light-hearted, joyful, exciting. Instead, I was sitting on a boat, delving into the abyss of my own emotions in response to a wild, injured dolphin. It seemed implausible that I could feel such an affinity to this mammal so quickly. But I did.

Back then, I didn't know just how important my connection to this dolphin would become. It would release me and teach me. It would help me find inner peace and a connection to the natural world. I would finally understand and revel in the simplicity of having fun, living in the moment.

This dolphin, whose name was Jock, became my sanctuary, my saviour. He and his world would heal me, teach me so many lessons, and show me that love was truly possible.

This is our story.

1

ANOTHER TIME AND PLACE

WHEN I FIRST jump into the water with Jock, it's the most remarkable experience I've ever had. His skin feels like cool satin against mine, his sonar clicks seem to vibrate through my body. He's checking me out, wondering what I am. The only way I can describe it is that I feel like he's exploring my essence. When I look into his eyes, I can see intelligence, curiosity and what I think is a twinkle of mischievousness. I've never been so close to a dolphin, or any wild creature, before. Sure, I've had pets that I loved, and I imagine they felt affection for me, but this is different. This mammal is demanding nothing of me apart from my attention. I am here, in his element, on his terms. I am interacting with a wild being, whose only reason for staying with me is because he wants to. There is no other enticement, no free fish, no coercion. And I feel there is no hesitation. I revel in the freedom of the moment and sheer wonder of it.

I know, however, that I am an intruder in his world. While I am seeking acceptance from him, it's his choice whether or not he will welcome me. As a human, I am supposedly a member of the most intelligent species on Earth, yet I feel humbled and insignificant in this dolphin's presence. I know little about his world, his kind, his needs and his thoughts. I really don't know anything about him. But what I do know is there's nowhere on Earth I would rather be in this moment.

Surprisingly, I'm not afraid. The water is deep and cloudy. Since childhood, I have been afraid of sharks. With Jock, however, I know I have no reason to be fearful. That innate knowledge is something that, to this day, I can't explain. Instead, I am in awe of how natural and easy it is to be in his company.

Tears well up. I don't deserve this trust. I am a human; not all humans can be trusted. At the edge of my consciousness, I feel a niggling fear. I push it away, but it remains a lurking anxiety during my long friendship with Jock.

After several minutes, Jock moves away. He swims a few metres then circles back. Reaching me, he pushes his snout into my right hand, which I'm moving backwards and forwards to keep myself stable. I'm close to the research boat, so I reach out with my left hand to hang on to the ladder. Jock's snout remains in my hand, his body hanging effortlessly in the water, keeping in place with just the slightest movement of his tail fluke. I begin to explore his snout, running my hand over it. I'm shocked at how rough it feels, quite unlike his satin body. His snout feels like sandpaper and his mouth feels lumpy and scarred.

LATER, I DISCOVERED these scars were the result of fishing hooks lodged in his gums. His snout was abrasive due to constantly rummaging in coarse silt and sand to dislodge tasty snacks, such as clams and crustaceans, from the riverbed. Unlike other dolphins, he fished alone. I imagined that might limit what kind of food he could catch.

Jock had other physical scars, too. His mangled dorsal fin had been disfigured from multiple entanglements in discarded nets and fishing lines, which had cut deep into the tissue. Where the wounds had healed, his fin was criss-crossed with bands of white scars. The tip of his dorsal fin seemed to be perched in a precarious position. How long would it be before it fell off, I wondered. Days, months? How could we humans cause such mutilation and pain through our carelessness? He also had marks on his body, old wounds from other injuries.

What physical battles had Jock been through? Had they caused him what we humans call emotional scars? Had he struggled and, if he had, how had he got through? How had he survived his battles to be here with me now?

Today, we understand more about contact with dolphins, and physical touch is discouraged. This is because relationships with humans aren't always good for them. Globally, there have been reports of people intentionally hurting or even killing friendly dolphins that seek out human companionship. There have even been anecdotal accounts of people shoving foreign objects and liquid down dolphins' blowholes. Other dolphins have died from being entangled in fishing nets or line, or being hit and killed by boats. In 2021, a friendly, solitary dolphin called Nick, who lived off the UK's Cornish coast and played with children and paddleboarders, was struck and killed by a boat.

While we never fed Jock, feeding dolphins is also discouraged, because they come to rely on humans for food and are at risk of entanglements. Times have changed since our research team swam so regularly with Jock. In retrospect, I feel incredibly lucky to have had such an intimate, close relationship with him; his injuries and scars helped me to reflect on my own childhood in a small town in Cornwall.

ANOTHER TIME, ANOTHER PLACE, yet once again I was close to water. I lived in Saltash on the River Tamar, just across from Plymouth, England.

Our house was an old, whitewashed miner's cottage. There were open fireplaces in most rooms, which my father would load up with coal, but it never felt warm. No matter how many fires were burning, the chill lingered.

For the first seven years of my life, I lived there with my parents, two sisters and a brother, before some of the family moved to Australia.

The house was halfway up a steep hill. At the bottom was a pebbled beach with stones of all shapes and sizes in various rusty hues. On a patch of grass beside the beach, I would sit for hours on a set of swings. It was a place of escape for me, and my sisters and brother. The beach was littered with small wooden fishing boats, their once-vibrant colours now subdued by time and sea air.

Sometimes, I would fantasise about whether I could turn one of them into a pretty sailboat, like those I occasionally watched from shore. I wondered if I could steal a bed sheet from the washing line and turn it into a sail, so the four of us could drift to the faraway lands I'd only ever seen flickering in grey tones on a small screen in the lounge room.

Occasionally, I waded knee-deep into the water, unconcerned about the coldness numbing my muscles. Staring at the horizon, I ached to know what was on the other side and what amazing sea creatures I'd meet along the way.

My brother, Mark, and I sometimes managed to lift one of the wooden boats and snuggle beneath it. Sunlight filtered through the slatted sides. We pretended we were smugglers, hiding out in a cave, avoiding capture by evil overlords. Sometimes Mark and I would stand on the shore and skim pebbles across the water. He always managed to make his skip gaily, while mine usually leapt once before sinking without a sound.

Another place for me to hide was our garden, which comprised a large parcel of overgrown land, full of weeds and crevices. My mother tamed some of it into a vegetable patch surrounded by roses and other beautiful flowers. I used to watch her weed. Occasionally, she dug up old light bulbs and other fittings. She told me my grandmother on my father's side had been a bit crazy, because she thought flowers would grow out of those bulbs.

I often escaped into some of the long undergrowth and crevices where I thought no one would find me. I imagined I could live there forever if I had to.

Up the hill, there was a sweet shop. When my mother and I walked there, I would look up at the Royal Albert Bridge spanning the Tamar. One day, I thought, I could catch a train across that bridge and never return. Our town wasn't far from the big city of Plymouth. We rarely went there, but my father visited it for work.

My father, Neil, was a good-looking man. What he lacked in height, he made up for in his unusual combination of features. He had a mass of wavy auburn hair. His permanent light tan was the result of his combined Sri Lankan and Cornish parentage. His eyes, the colour of unbaked clay, were striking against his skin. He had a broad mouth, which rarely formed into a smile. His round face featured a deep frown line, from the middle of his forehead to the top of his wide nose.

He wore the same tea-brown belt every day, which had a little yellow plastic box containing a measuring tape clipped onto it. He prided himself on his intellect and engineering prowess. According to him, he could fix or build anything.

The only time I saw him take off his belt was when he was about to use it on my sisters or brother.

My mother, Doreen, was a handsome woman, and she knew it. She took pride in her hair, which was thick, dark and luxurious. It tumbled to her shoulders in waves, created by foam rollers she wore overnight.

Her long face had generous cheekbones and a wide mouth with plump lips. When she smiled, it was entrancing. She had brown eyes, which she often enhanced with thick black eyeliner. They were slightly darker than my father's eyes. Compared with my father's skin, hers was light, milky and unlined. She was taller than he, and fourteen years his junior.

My mother was a vain woman, and I would often catch her examining her reflection in the dressing-table mirror, turning left and right, staring at her slim waist and round, yet perfectly proportioned hips and breasts. She often complained loudly about the unsightly varicose veins in her legs, which she rarely showed to the world. She said they were the result of carrying us children.

To me, however, even in her dressing gown and rollers, her face slathered with Pond's Cold Cream, she always looked beautiful.

My sisters were twins who looked nothing alike. Dot was fair, tall and slim. Her thick, straight strawberry blonde hair stopped just below her shoulder blades. Under her heavy fringe sat large aquamarine eyes. Ana was short with a milk-coffee complexion similar to my father's. Her dark unruly curls, which she hated, bobbed on her shoulders. The only thing the twins had in common was their ocean-coloured eyes. Even their personalities seemed opposite. They were nine years older than I was.

My brother Mark was a mix of the two. Five years older than me, he was lanky with straight brown hair. His soft cocoa eyes, rimmed with long dark lashes, always seemed to be on the edge of tears. His skin had a slight hue, not as dark as Ana's. He had long, skinny legs with knobbly knees, which often appeared about to give way. He never looked comfortable, even when he was lying on the couch. He seemed awkward, ready to leap and run like a deer at the slightest danger. He was also shy and often wet the bed. I loved him, even though he seemed different from me in almost every way.

I had fine, straight, shoulder-length brown hair, which was peppered with auburn streaks only visible in daylight. Like Mark, my skin was not as dark as Ana's, but I was not fair. Like my brother, I was skinny. People often commented on my chocolate-brown, large eyes. I hated that they were darker than anyone else's. When I looked in the mirror, they looked almost black. From reading fairy tales, I knew only witches had black eyes.

So, I called myself Princess. It fitted with the imaginary world I visited every night, lying in bed, often worrying about the noises from my parents' room next door.

THE YEAR WAS 1973. I must have been about five. It's my earliest memory. My mother and father were kneeling at opposite ends of our sprawling garden with their arms outstretched. I was standing in the middle.

"Melody, come to Mum," she called.

I hesitated.

"Melody, come to Father," he yelled.

I was confused. I didn't know what to do. Were they playing a game?

I looked back and forth at them. Each was calling my name, smiling and urging me towards them. What was the right thing to do? I felt tears on my cheeks.

I ran to my mother.

Sweeping me up in her arms, she carried me over to my father.

"I told you, she loves me more," I recall her saying.

"Well, I'm done with her," he said, and walked back to the house.

Later, I heard them screaming about it in the kitchen. Maybe I'd made the wrong choice. Doors slammed. One of my sisters ran out of the front door. It was going to be another bad night.

ALTHOUGH THE HOUSE was comfortable enough, I always felt a sense of unease. It was like all of us were holding our collective breaths, waiting for the next explosion. It seemed strange to me that, after each flare-up, an uneasy peace settled for a few days as if nothing had happened. Inside the house, it always felt like a storm was brewing. The air became increasingly thick and heavy with threatening energy. I found myself on the lookout for any telltale signs about when the storm would break—sometimes, it was just a sideways glance between my mother and father.

During one of these build-ups, my sister Dot didn't come home for dinner. This was a problem. I could feel the growing tension around the small wooden kitchen table, where the rest of us sat on mismatched chairs. Everyone was silent. Mark played sullenly with his food, pushing bits of carrots and turnip around, shooting glances at my father, who sat next to him. Ana spooned the food mechanically into her mouth, elbow on the table, head resting on her hand. She seemed lost in thought.

My father sat opposite my mother. He stabbed aggressively at his plate of stew.

"She's out with that boy again," he glared at my mother. "There'll be hell to pay when she gets home."

Mum ignored him and picked at her meal.

"She's a bloody whore, just like you. Like mother, like daughter." His face was starting to redden.

He thrust his spoon in my direction. "And this other daughter has got the same gene. She'll grow up to be a useless slut, just like her mother."

While I didn't know what a slut was, I gathered it wasn't good. But if I grew up to be like Dot, I thought, that would be okay with me.

My mother stopped playing with her food and looked up at him. Her eyes narrowed into slits. "You're such a bastard."

She stood and picked up her plate of stew. I grabbed the undersides of my chair, bracing for what I knew was coming. My brother scampered out of the kitchen. I heard his bedroom door slam.

My father's lips quivered into a sneer. "You'd screw anyone."

My mother hurled her plate at the floor. The china shattered, splattering bits of vegetables and meat onto the slate tiles. Gravy settled between the joins.

She leaned over, planting her arms on the table. She pushed her face forward until it was centimetres from his. I could see the veins in her neck. She grinned mockingly. My father jumped up and pushed her shoulders hard. She stumbled back, losing her footing, and thumped onto the floor, landing in some of the deconstructed stew.

Ana leaped from her seat. "Stop it, you two! You're like bloody children."

They ignored her and stared at one another, my father standing, my mother on the floor. The air felt as thick as the gravy splattered across the floor.

Ana ran from the table. I sat, not knowing what to do. I wanted to cry but experience told me that would only make things worse.

My mother grabbed the edge of the table and pulled herself up. Ugly brown stains discoloured her pretty powder-blue dressing gown.

My father's frown line had deepened. His face looked like it would split in two. He pointed at me. "She isn't even mine." There was foam at the corner of his mouth. "If she was mine, she wouldn't have gone to you."

"She is your child," my mother hissed back at him. "You know damn well she was a mistake. You raped me, and look what happened. You're such a bastard. I hate you."

I sat, frozen with fear, longing to escape. I knew I wasn't planned. I was a mistake. I'd heard that before. My mum also told me that my dad had wanted her to get rid of me, but she chose to have me.

At that moment, I wished I was bigger, so I could knock them both senseless. I couldn't hold the tears back. I started sobbing loudly.

"I didn't ask to be born," I shouted at my mother. "I don't want to be here."

She turned to me, her face flushed.

"What a stupid thing to say. No one asks to be born." She said this with such venom, I couldn't believe she was the same mother who, earlier that day, had taken me in her arms and told me she loved me.

My father's hands were balled into fists on the table. My mother walked over to him and slapped his face. Then the two of them were wrestling, grunting, shrieking, swearing, limbs entwined, clumps of hair in hands, fists flying.

Suddenly, the front door banged, and Dot walked into the room. She threw back her head and screamed.

I couldn't control my sobbing. I was gulping for air. My heartbeat couldn't keep up.

My father pushed my mother away. She spun, her face hitting the corner of an upper kitchen cupboard. She sank to the floor, and slumped against the bottom cabinet.

My father strode across to Dot. He grabbed her strawberry blonde hair and pulled it so hard, I thought her neck was going to snap. She was crying and her eyes were shut tightly.

"You've been out whoring, haven't you?"

"I just went to see Pat." She could barely get the words out. I saw her knees trembling. He looked like he was holding her up by her hair. "You're hurting me!"

"You want to know what hurt is?" he yelled. "Here's hurt."

As he let go of her hair, she collapsed to the floor. He removed his belt and struck her hard with it. She curled into a ball, begging him to stop. He brought the leather belt down on her, again and again.

My mother sat still, eyes glazed. Her voice was slurred. "That's enough, Neil. Let her go."

My father stopped and turned towards me. I felt every muscle in my body tense, fearing I'd be the next to feel the lash of his belt. My streams of snot and tears had saturated the top of my yellow turtleneck.

He rethreaded his belt and strode past me, out of the kitchen. My heart was still pounding but I felt the fear fade.

Dot slowly stood up and looked at my mother. Her face was blotchy and swollen. "I hate you. I can't wait to leave. One day, I'm going to marry Pat and we're going to run off together."

"Do what you want, Dot." My mother seemed devoid of energy. She sounded tired, resigned. "I'm getting a migraine."

My sister ran out, slamming her bedroom door.

My mother turned towards me. "I love you, Mel. You're my Princess."

I knew how much I wanted to believe her. "I love you too, Mum."

"Now, go to bed, and put that sweater in the washing basket. I'll come and tuck you in."

Our black cat Puss was sitting on the sill outside the kitchen window. I opened the door and he slunk in. I picked him up. He was all fluff. I carried him to my room, changed into my nightie and sat on the bed, waiting for Mum.

Tomorrow, I told Puss, we'd go hide in the bushes and never come out.

GRIM DAYS AND TENSE, terrifying nights passed. It was always the same pattern. A big fight was followed by days, sometimes weeks, of uneasy calm. My father went to work, my mother looked after us and sometimes went to work at a jam factory. When my father was at work, she could be so loving towards us. She'd take us to the sweet shop, hug us fiercely and tell us how much she loved us. I couldn't understand how she could turn into something so different so quickly. I yearned for this caring mum, the one I could cuddle, who called me Princess, told the twins how fashionable and beautiful they were, and even laughed at Mark's jokes. Mum often blamed her dark moods on her high blood pressure. I didn't know what that was, but I didn't like the symptoms.

When my mum was in one of her kind, generous moods, she tolerated my father more easily. He'd come home from work, sit down for dinner and they'd talk to each other without yelling. Although the air still felt heavy, it was better than another screaming match or fistfight. We kids didn't say much. As my father often reminded us: children should be seen and not heard. While I had lots I wanted to say, I thought it best to keep my mouth shut.

On a few occasions, my father took my brother and me fishing down at the shore. I always dreaded it. We carried small containers filled with fat earthworms from the garden. He carried his fishing rod as if it was a military baton. We followed him down the hill in silence, staying a few steps behind as he strode ahead. He never held our hand. Our job was to put the worms on the hooks. I could never bring myself to pierce the helpless, squirming little animals with the razor-sharp hook. My father sighed, took the worm from my hand and threaded it himself. Mark and I sat in silence on the beach as he waded into the water, cast his line and waited for a bite. Occasionally, he landed a mackerel. He whacked it down hard onto the pebbles and gutted it. The gulls seemed to be the only ones enjoying the experience. He muttered how none of us should ever take a free meal for granted, and told us we'd have to learn how to fish to earn our keep. Although I never enjoyed them, these rare trips were some of the few occasions I can remember spending time with my father. A morning's fishing usually put him in a good mood, which meant a row at home that day was less likely. So, in my mind, it was worth it.

The strained peace never lasted. Before long, the tension in the house rose again, the air tingling with an electric current that would eventually lead to a blown fuse.

One particular week, I would have been perhaps six years old, my mother seemed more irritable than usual. The arguments increased until my parents were yelling at each other every day. There wasn't a break. The screaming, banging on walls and sobbing seemed never-ending. I longed for it to stop. Night after night, I heard angry voices, shrieks and loud thumps coming from their room. No matter how tightly I covered my ears with pillows, I couldn't block out the sound. I didn't know what they were fighting about, but it sounded like my father was accusing her of something and she was denying it.

This seemed to go on for weeks. Both of them carried more bruises.

Every morning, Mum checked Mark's bed. He was wetting it more often. This infuriated my father. Depending on her findings, punishment followed. The belt came out. My father screamed at Mark, calling him weak and stupid. Later, I heard my brother sobbing in his room. I felt so sorry for him. He occasionally missed school. I was so glad I didn't wet my bed.

Mark spent even more time in his room; I rarely saw him. When I did, I sometimes managed to persuade him to play hide and seek with me in the garden. I didn't blame him for staying in his room. Usually, he only appeared to go to school and at dinnertime. He always looked pale. He seemed terrified of my father and the never-ending fights.

My sisters also made themselves scarce. They left the house as often as they could. I wished I were older, so I could go out by myself. When I could, I spent time in my garden nooks, finding a spot where I could sit and read comfortably. I lost myself in stories of majestic castles, hidden underground hallways, wardrobes leading to enchanted worlds, tea parties with grinning cats and secret gardens where rabbits talked.

One morning, my father didn't go to work.

He brought the Morris out of the shed and parked it outside our house. This was unusual because we didn't use the car much. Dad mostly took trains.

"Are you ready?" he asked my mother, who was wearing a green coat, her hair perfectly set. She was rifling through a small, battered brown case on the kitchen table. I hadn't seen that case before. Its stitching was coming undone, the latches were tarnished, it had a hole in one corner and was badly scuffed on the sides. It seemed to represent my family life perfectly.

Dressed for school, Dot and Ana were standing in the kitchen, watching her.

"I'll be out in a minute," Mum said. My father walked out the front door.

The three of us watched as she secured the sullied locks.

"Where are you going?" I asked.

She glanced at me, grabbed the bag and walked out. We followed.

"Get in the bloody car," said my father through the driver's window.

My mother stopped, dropped the shabby case, put her face in her hands and started to cry.

"For Christ's sake, stop your bloody whining and get in the car." My father's face started to redden.

My mother stepped off the curb and stood in the middle of the road.

"What the hell are you doing?" my father yelled.

She ignored him and kept standing, as if in a trance, in the middle of the road. Moments later, she dropped to her knees and lay down. She folded her arms across her chest and started rolling down the hill.

"What the hell?" Dot shouted.

"Mum!" I cried.

"Mum, get off the road," Ana yelled.

We watched as she gained momentum. My sisters started running after her. I followed. My father stayed in the car, craning his head out of the window.

It was a steep hill and she was rolling quickly. The road was uneven, which caused her body to move erratically. I was so worried about cars.

When she reached the bottom where the road flattened out, she just lay there with her eyes closed, arms crossed and hair in disarray. Little bits of black stone clung to her green coat. I thought she was dead.

A few seconds later, my father ran up beside us, puffing. He grabbed her arms and pulled her up.

"You're bloody insane," he said to her closed eyes. "Now, get back to the goddamn car and stop being such a crazy bitch."

A few of the neighbours had come out and were standing in their front gardens, watching.

My mother stumbled as she gained her footing. Looking groggy, in a daze, she slowly walked back uphill.

At the car, my father snatched her case from the pavement and threw it into the back seat. He got into the driver's side. Mum eased into the passenger seat. She wound down the window. She sat quietly, looking dishevelled. We gathered round. I put my hands on the window.

"Where are you going?" I asked.

She looked at me, fumbled in her pocket and drew out her scalloped handkerchief. "Keep this to remind yourself of me," she whispered. "I don't know when I will be back. I may never come back."

I took the hanky. It smelled like lavender. "No," I sobbed. "You can't leave me with Dad."

In the driver's seat, my father was staring straight ahead, emotionless, silent. I heard the rattle of the engine, smelled the car fumes.

"Sorry, Princess. Your sisters will look after you." She started crying.

I clung to the window, my knuckles turning pale. I felt arms around my waist. One of my sisters tugged at me, trying to pull me away.

"No!" My head was spinning. "No, don't leave me!" I tried to hold on, but the pull was too strong. I let go.

The car pulled away. The three of us watched, holding hands, sniffling. I couldn't believe it. I saw my mother's face appear out of the window as the car drove uphill. She waved.

"It's okay, Mel," Ana said bravely. "She'll be back."

Ana was right. A few days later, Mum returned. She behaved as if nothing had happened, and never mentioned the roll down the hill or her teary farewell again.

While she was away, I felt abandoned. I avoided my father whenever I could. I had nightmares and frequently woke up, my heart pounding, feeling like I needed to run.

As I lay awake, I listened to the creaks and groans of the old cottage. Puss was the only witness to my night terrors.

2

—

A DAY ON
DARTMOOR

I BEGAN STUDYING with Mike in 1990 but his first encounter with Jock had been two years earlier, in 1988. He along with fisheries officers tried to catch Jock after a report that a dolphin with a deformed dorsal fin had been spotted with a spear in its side.

I learned that Mike and others had watched him for a while but were unable to catch him. Jock had managed to dislodge the flounder spear, but the scars from the weapon remained. I later found out that previously, Jock had also been rescued from entanglement in a discarded net. I also learned he'd originally been named Jock by a local photographer as he lived in an area of the river called the "Angas Inlet." Apparently, it had seemed fitting that this solitary dolphin have a Scottish name too.

Mike told me that, after watching Jock for a year, during which he'd again became entangled in fishing line, it was clear Jock was a solitary. But why? Was it his deformed fin that made other dolphins ignore him, shun him? This theory was put to rest when researchers saw other dolphins swim into the inlet and approach Jock in a friendly way. He swam with them back towards the main river. For some reason, however, he then stopped and refused to go with them, remaining in the inlet. Did he stay there because he had suffered and was afraid of leaving his environment, or did he just feel comfortable on his own in the warm water of the inlet?

What other reason could there have been? Male bottlenose dolphins tend to pair off, Mike said, so maybe Jock had once had a friend, and maybe he'd died. Or perhaps it was simply that his mother had died in this part of the river, and he had been orphaned and reluctant to leave. Was he still waiting, hoping she'd come back?

We will never know the reason.

But we do know that fishing lines, boat strikes, pollution and malicious attacks by humans are constant hazards for dolphins living in the river.

As I begin the research trips, I learn more about the Port River, which is close to Adelaide, in southern Australia. It is a city of more than a million people. The Port River is an industrial and trading hub, and has historically been impacted by discharges from factories, fuel waste, stormwater, wastewater and other pollutants. Its shores are dotted with industrial facilities such as grain terminals, cement-works and shipyards.

Despite this, the Port River is home to Jock and dozens of other resident dolphins. The area where Jock likes to hang out is artificially warmed by water discharged from the Torrens Island Power Station, one of two on the river that generate electricity for Adelaide.

The river is located halfway up Gulf St. Vincent. The Gulf is home to other bottlenose dolphins, common dolphins and an abundance of diverse marine life. Its shallow waters are enjoyed by swimmers, anglers and boaters.

Even though the Port River faces challenges from human activity, parts of it remain breathtakingly beautiful, particularly the ancient mangrove forests lining the shores near Jock's home.

The mangroves provide vital habitats for young fish and many other species, which in turn create a food source for dolphins and other marine life. Their quiet channels are also the perfect place to hang out with a lonely, wild dolphin.

JOCK HAS BEEN *physically interacting with Mike and other members of the research team for some time before my first visit to the mangrove channels, initially playing games with a stick or paddle, and gradually letting himself be touched. Mike is cautious about who he allows into Jock's world. He wants to keep a lid on the number of people who know about this friendly dolphin, for Jock's own protection.*

I consider myself extremely lucky that I have been trusted and can get to know him so well. It's clear that humans, including myself, fill a gap for Jock. For me, however, Jock is filling a gaping hole in my own heart.

Those days I spend in the water with a playful creature are unforgettable. So different from the stress and tension of my early family life. They reminded me of how rarely I'd been able to escape as a child and just have fun. I did, however, have some days of real enjoyment, although they were few and far between.

THE PATTERN OF FIGHTS followed by stretches of uneasy peace dotted with loving devotion from Mum continued. As the twins spent even more time going out with their friends, Mark and I grew closer.

As I approached my seventh birthday, the twins both had boyfriends, which my father didn't approve of. I saw less of them and, when I did see them, they seemed more grown up. They also seemed braver, not so afraid of my father; they'd often yell back at him. They also began standing up to my mother, which led to more arguments. I tried to zone out, escape into nooks and books, or visit the beach with Mark.

Despite how unhappy I was at home, I felt lucky to live in a cottage with such a big garden. Many houses in the village were small, grey and crammed together along narrow streets with little or no garden. Everything in Cornwall seemed damp, even when the sun came out. Sometimes a salty mist rolled in from the sea and the whole village was hugged by cloud. It was magical to watch the pebbled beach disappear into white in the morning, only to reappear in the afternoon.

So, too, were my occasional trips to the country.

One morning, not long after my mother had returned from her mysterious trip, I got up early and dressed in my favourite outfit—a yellow turtleneck and a bright-red knitted vest flecked with silver. The sparkly threads in the vest had a regal quality, which I adored.

My pants were pale blue with tartan cuffs. I was a fan of the Bay City Rollers and my mother had made them for me. I grabbed my scarlet coat, which reminded me of Little Red Riding Hood. When I wore this ensemble, the world seemed brighter. The vibrant colours lifted my soul from my drab, grey surroundings.

I heard a horn honk.

"Come on, Mel!" Ana called from the kitchen. "Neil is here." I could hear the excitement in her voice.

I was excited, too. My sisters and their boyfriends some-times took me on weekend drives to the country—and I had been looking forward to this all week. I was going to

Dartmoor with Ana and Neil for a picnic. Dartmoor was wild, lush and rich. Even though it was little more than an hour's drive from the cottage, it felt a million miles away.

I ran from my bedroom and joined Ana in the kitchen. I grabbed her hand.

My mother was at the sink, washing dishes.

"Don't stay out too late," she said to Ana. "You know what your father's like. And keep an eye on her." She pointed a wet rubber-gloved finger towards me.

"Okay," Ana said. Opening the fridge, she took out a brown paper bag filled with cheese-and-chutney sandwiches made the night before.

We ran outside to where Neil's car was parked on the curb. It was a small, dark green, two-door van with seats in the front and a banged-up, rusting metal floor in the back. Bits of plumbing equipment were piled in the rear corner. A blue tartan rug covered the rest of the floor. The rug had seen better days—it was torn, tatty and matted, with oily splotches. It would be my seat for the journey, but I was happy.

"Hello, beautiful." Neil leaned out of the window to kiss Ana quickly on the lips. He turned to me. "Hi, Mel."

Ana tilted the front seat forward and I scampered into the back.

"Sorry Mel, got a bit of kit back there," said Neil, in his thick Cornish twang. "Had a late job last night and couldn't be bothered to clean out the van."

I leaned forward between the two front seats. "It's okay, thanks Neil."

For as long as I could remember, Neil had been Ana's friend. He had crooked, slightly bucked, stained front teeth and a long, skinny face. His lank brown hair reached his shoulders. He wasn't particularly handsome, but he had kind copper-coloured eyes and I liked him.

We set off, leaving the narrow streets of Saltash behind, crossed the River Tamar and headed for the moors.

The rear of the van had small rectangular windows on either side. I stared out at the never-ending lush green fields, dotted with cows and sheep. There weren't many other cars on the road. I snuck closer to the front to listen to Neil and Ana talking.

"It's just, sometimes, I'd like to go for a drive in the country without your little sister tagging along."

"Neil, come on. You know what it's like at home. It's bloody awful. I've gotta get her out of that house. She loves coming with us."

"Yeah, well, your family's bloody nuts. Your mum went mental on me last weekend just for getting you home a bit late. I'm just saying, it'd be nice to have some time alone on weekends. I don't mind sometimes, Ana, I really don't, but we need some privacy."

"Okay, we're going to the stream, right? When we get there, I'll ask her to go and see the horses. I stole some carrots. I can keep an ear out for her. She's sensible. We can talk and ... whatever."

She giggled, reaching over to tickle him.

"Oy, watch it, girl!" He smiled back. "Gotta keep my eyes on the road."

I stared at the passing fields. I knew my sisters' boyfriends didn't want me always tagging along. I was still only six and three quarters, while my sisters were sixteen. I didn't want to be a burden, but I loved these escapes to the country.

We turned off the main road into a lane, barely wider than a car. I wondered what would happen if someone was coming the other way. The high hedgerows on either side made it seem like we were driving inside a maze, like the ones in the grand castle gardens in fairy tales.

We turned onto a bumpy dirt road. I bobbed around in the back of the van. Neil's plumbing gear clanged in the corner.

"Sorry, Mel. Bit rough 'ere."

After a few minutes, he stopped the van. We got out. I heard familiar tinkling and burbling coming from a narrow stream nearby. I'd been to this spot before.

The water sliced through the long grass, tumbling over round grey rocks jutting out of the shallow bed. It was crystal clear, unlike the water from the taps at home. The air smelled of earth scented with wildflowers. Dappled light played on the grass from sunlight streaming through the trees. Although the breeze had a chilly edge, it felt warm. I raised my face to the glorious rays and the world suddenly became illuminated. It felt good. I breathed in deeply.

Neil grabbed the rug from the back of the van and laid it on the grass.

"Here y'are, Mel." Ana put her hand into her jacket pocket and pulled out several small carrots. "I saw some horses just over there as we drove in. Why don't you go see if you can find 'em? They'd like a treat. Yell out if you get into trouble."

I was so excited. "I'll be okay. Thanks, Ana!" I stuffed the carrots in my coat pocket.

Neil and Ana settled onto the rug and I ran off in search of the ponies. I'd seen some on previous trips but had never approached them.

Dartmoor is famous for its wild ponies, which roam the countryside. The breed has been there for thousands of .

years. Most of them are stocky with woolly coats, which protect them from the biting cold and wind.

As I walked through the long grass, I looked around. Huge, grey mottled rocks protruded from the undulating terrain like the fingers of underground giants reaching for the surface. The landscape felt really old.

The hills looked like they'd been draped in patchwork quilts knitted with balls of multi-hued green wool. While some of the land had been tamed, other parts remained lush, wild and overgrown. I felt free, exhilarated. The breeze tousled my hair and stung my eyes, but I didn't care.

Up ahead, I saw three ponies, their woolly coats ruffling in the wind. They looked up from their grazing and eyed me warily. Unlike my mother, who feared horses, I thought they were magnificent, majestic and gentle. Their size didn't intimidate me. Slowly, I approached them. Their ears flicked.

"Hello, I'm not going to hurt you."

Two turned their backs and trotted away, but one stood still, eyes locked on me.

"Come on, I've got some carrots." I held one out.

The pony was a deep biscuit-brown, with a luxurious ebony mane falling over his eyes. He was stout, with delicate feathers of fur cascading over his hooves. His brown eyes, lined with the blackest, longest lashes I'd ever seen, didn't leave me. He stepped forward.

I held my arm outstretched, carrot in hand.

He approached my hand, sniffed and snorted. I could feel his warm breath on my skin, and saw the mist blow from his nose. He moved his lips over the carrot and took it gently.

Crunch, crunch. It was gone. He sniffed my hand. I withdrew another carrot from my coat and offered it. He took that one, and a few more. It was a magical moment. I wanted it to last but I'd run out of carrots.

"See you next time. I'll find you again," I promised.

As if on cue, he turned and sauntered away. His large bottom swayed and his long tail flicked as he returned to the herd. I thought it was about time I went back to mine.

As I approached, I saw Ana and Neil entwined on the rug. I decided to make some noise by stomping and rustling the grass.

They looked up.

"Hiya, Mel. Did you find the horses?" Ana asked. Her face was flushed, her curly hair in disarray.

"Yep, one of them let me feed him. I don't know if it was a him, but it looked like a boy. He's chubby."

They both laughed.

"Okay, time for lunch," said Ana.

I joined them on the rug. I could hear the stream gurgling, the grass rustling, the occasional chirping of a bird. Nature was amazing, I decided, as I chomped into the cheese sandwich. I felt so at peace.

We stayed, lying on the rug, listening to the sounds of the land, sipping tea from a thermos.

"Righto," said Neil. "Time to get you back, Mel. Ana and I have a movie to get to."

Reluctantly, I got up and got in the van. On the drive home, Neil and Ana murmured in the front. I thought it better not to eavesdrop. When I got home, I decided to immerse myself in one of my favourite books.

As soon as we arrived at the cottage, I ran inside, called to my mother that I was home and grabbed my book from under my pillow. I wandered outside and settled into my favourite nook. It was mid-afternoon—the light was weakening but strong enough to filter into my hideaway, allowing me to read.

The paperback was tatty and dog-eared, but the cover image was vivid. A lion playing with three children. I had read this book many times. I found my spot and was transported instantly to the magical land of Narnia.

Mr. Tumnus, the faun, was making toast for his young guests. It was snowing outside but it was warm in his living room. I could hear his knife scraping across the rough surface of the toast. I smelled the rich butter, saw it melt. I was mesmerised by the amber glow from a gas lamp in the corner. I felt cosy and safe.

I stayed in my hideaway until the light faded and I grudgingly returned to the real world. The fun was not to last.

A DARK STAIN, the colour of crushed violets, dominated my mother's left cheek. Her lower eyelid was puffy and a deep rose pink. The hue reminded me of the geraniums that grew wild in our front garden. I'd seen parts of her spotted with similar colours before, often dark crimson and purple. Sometimes, I'd seen her with what my brother called a "shiner." On occasion, I saw my father with similar ugly marks on his face and arms.

I overheard my mother speaking on the phone. She worked part-time in a jam factory and was calling in sick.

"I'm going to walk you to school," she said to me, relieving my sisters and brother of the duty. "I just need to cover this up."

I had no idea where my father was. He was probably at work. He was a sales engineer and, according to him, a talented one. His absence didn't surprise me. Even when he was home, he rarely talked to me unless in anger or frustration.

He looked over me, or through me. He slammed doors on me. I can't remember a time he hugged me or held my hand. Often, I felt like an inconvenient boarder, who he had to put up with and navigate around. It seemed clear to me that I simply wasn't worth bothering with. He didn't want to waste his precious time or superior intellect on an insignificant irritant unless, of course, he wanted me to thread worms on a hook. The only plus was that he didn't hit me with his belt, like he did the others—although I often worried I would cop it one day.

My mother emerged from her room. The make-up she'd slathered on only seemed to make the bruising more obvious. I didn't say anything.

We stepped out and started walking. The school was roughly a ten-minute walk from our house. It was cool and overcast, with a slight breeze. The air was laced with salt and a whiff of decaying vegetation. It wasn't unpleasant. These were familiar smells. No matter the weather, I wore the same bright clothes I'd worn on the trip to Dartmoor. When it was cold, I always wore my bright Little Red Riding Hood coat.

"Hello, Mrs. Horrill."

We stopped in front of a whitewashed cottage like ours, a couple of doors down. My friend Sissy lived here. She was two years older than I was, with an abundance of deep russet hair and freckles sprinkled across alabaster skin. She giggled a lot. We often played together on the grass near the pebble beach. Sissy liked to pretend she was a pop star, and would sing and swing while I swung in unison and listened.

"Mrs. Browning. Good morning," my mother said in what I called her "posh voice." Her tone changed. Her words seemed rounder. She sounded like the Queen.

"Seaweed's a bit whiffy this morning, ain't it?" Mrs. Browning drawled in her Cornish accent. She had rollers in her hair and a cigarette hanging from the side of her mouth. I could see her looking intently at the left side of my mother's face.

"Yes, well, this is why I always carry a handkerchief with a little lavender oil on it."

My mother pulled out the white cotton hanky with the scalloped edge from her jacket pocket. She held it up to her nose and inhaled deeply. "Lavender also does wonders for one's nerves."

Mrs. Browning kept studying my mother's face.

"You know, we hear things coming from your house. None of my business, of course. I just want you to know, Melody is welcome here whenever she wants."

My mother lifted her head and sniffed. I'd seen this move before—it often preceded a dressing-down or temper tantrum.

"Why don't you just leave, Doreen?" Mrs. Browning continued, her voice softening. I was surprised she'd used my mother's Christian name. My mother glared at her.

"And where am I going to go?" she snapped. "My father told me, I made my bed and I must lie in it. He won't help. My mother died when I was sixteen and I have four kids weighing me down."

Mrs. Browning stepped back.

"Thank you, but we're fine," my mother continued. "You're right, it's none of your concern. Good morning, Mrs. Browning."

As we walked on, I looked back. Mrs. Browning was staring after us. Our eyes met, then she looked down, turned and went inside.

We arrived at the school gate. I was late.

Mum kneeled in front of me.

"You know I love you, Mel."

I looked into her eyes. The left one looked even puffier now.

"I know I'm not the world's best mother, but I do love you and the kids. You're my little Princess."

"I wish we could all just leave," I said. "I'm sick of the screaming and fights and everyone getting hurt. You and Dad hate each other."

My mother looked away. "I can't afford to leave. But your father and I are talking about moving. It's still a while off though."

"Where?" I demanded. "Why is he coming?"

She didn't answer. She told me to have a nice day and behave, then walked away.

So many questions were swirling in my head. Were we leaving Cornwall? Were we going to the big city over the bridge, or maybe even to London?

Gee, was I off track!

ONE EVENING, MONTHS LATER, the six of us were eating fried eggs and chips at the kitchen table. It was one of the simmering peaceful periods. It was cold, the house smelled damp, rain was pelting on the tin roof and a fire was crackling in a corner of the kitchen. Puss was sitting in front of it, seemingly hypnotised by the flickering flames.

My father put down his cutlery and surveyed us.

"We're going to Australia," he said. "Not for a visit. To live."

We stopped eating and looked at him. I glanced at my mother, who was intently dunking a toast soldier into her egg yolk. She didn't look up.

"I've lined up a job there. We'll be leaving in a few months. We're going to a place called Adelaide. It's in southern Australia. We'll be spending a few weeks at a hostel for immigrants when we arrive. Then we'll get a house. It'll be a fresh start."

He picked up his fork and shoved a plump chip into his mouth.

The fire hissed and spat. Puss scampered off. Rain drummed on the roof. The room felt hot. I couldn't believe what he'd said. We weren't going to Australia, surely? This had to be joke.

"I'm not going back," Dot said sharply, throwing her utensils onto the table. "I'm not going to bloody Australia, especially not with you two."

"I don't want to go back either," Ana chimed in.

Mark and I looked at each other. My mother kept jabbing at her egg yolk.

"What do you mean 'back'?" I asked.

"We lived there for a while. Mark was born there. I'm sure Mum's told you this already," Ana said, turning to my mother.

Mum looked up from her ruptured yolk. She took a breath as if about to say something.

My father exploded.

"You WILL get on that bloody plane, even if I have to drag you by the hair!" The vein in his neck pulsed, colour began rising in his cheeks and his frown deepened.

"Like hell!" Dot's chair skidded on the floor as she pushed herself away from the table. She stormed out of the kitchen. Ana followed quietly.

I sat, disbelieving, in the tense silence. I didn't remember Mum telling me anything about their life in Australia before. Why were we going back? Why? It didn't seem to have worked the first time. I was determined to ask Mum again later, with my father out of the room.

Now, I felt depressed. Despite all the rows, this was my home. I loved meeting my friend Sissy at the beach, listening to her belt out Linda Ronstadt songs on the swings. Watching Mark skim pebbles across the water. Hiding under wooden boats. I hated the idea of leaving the garden with my secret nooks.

Australia was another world. I'd watched TV shows about the outback and kangaroos. I liked kangaroos, but I didn't want to live in the hot desert with them. Then there were the snakes and spiders. I didn't like spiders. Even the small daddy-long-legs I sometimes uncovered in the dusty corners of my room scared me.

Finally, I said, "Don't they have big spiders there?"

No one responded.

"This is not up for discussion." My father glared at me.

I turned back to my egg and chips, feeling confused and anxious. Mark had stopped eating. He kept his eyes on the table.

Later that evening, my mind whirling with questions, I sidled up to Mum as she was doing the dishes. My father was working in the garage. I grabbed a tea towel. My mother raised her eyebrows as she looked down warily. I rarely volunteered to do the cleaning up, unless I'd broken something or made a mess. I grabbed a plate from the rack and started wiping it.

"I don't remember you telling me about living in Australia before."

"Actually, I have mentioned it. But it's not something we really talk about."

"Why? What happened?"

"Some things should be left unsaid, Melody." She sighed. When she used my full name, I knew she was getting angry with me.

I decided to push my luck anyway.

"Where did you live? What was it like?"

She stopped scrubbing at bits of egg clinging to the cast-iron frying pan. She placed the pot in the sink, her gloved hands lingering on it in the soapy water. She stared ahead, looking at nothing.

"Look, your brother was born there, in Sydney. Your sisters were about four or five. We went there for your father's work for a few years. He built a house at a place called Blacktown."

Her face was emotionless, delivering the facts in a dull tone. It sounded like she was reeling off a shopping list.

"Your father did a good job of building that house. We had a cherry tree in the backyard. Then your father got transferred back here. We sold the house and came back."

"So, what was it like?" I pressed. "Sydney, what was it like?"

"Nice, actually, warmer than here. I quite liked Blacktown. You wouldn't have liked the spiders though."

She looked at me and smiled. Now, it seemed she was teasing me.

"The funnel-webs were huge. They built webs in the lawn. Your father used to go around the backyard, pouring hot water down the holes in the lawn where they burrowed. They were nasty—their bite could kill you. We also had huntsman spiders. They were as big as your hand, but they didn't hurt you. So we left them alone. They ate other insects."

I was horrified. I visualised a back lawn filled with large, black hairy spiders, their big white fangs emerging from underground tunnels, scuttling across the lawn towards the house. Ugh.

"That's all there is to tell, Mel." She picked up the pan and resumed her battle with the obstinate egg. "Just let the rest be. It's history."

I wasn't going to get any more information out of her and, given the spider story, I wasn't sure I wanted to. Mum's reluctance to talk about stuff that happened in the past didn't surprise me. She always said that things in the past should stay in the past.

I kept thinking about the funnel-web spiders. Over the following weeks, I became obsessed with finding out more about this parched, insect-ridden place with weird-looking animals. I borrowed reference books from the library. I learned it was a big island, a lot bigger than England. It was blisteringly hot in summer and yet, weirdly, people snow-skied there. The only appealing aspect was that it was surrounded by sea, but even that was filled with sharks and jellyfish that could kill you. I looked at images of vast white beaches. Unlike Saltash, the shores were sandy, not pebbly. Mark probably wouldn't be able to skim pebbles there.

I asked my sisters to tell me about Sydney a few times, but they never wanted to discuss it. I decided that, maybe, there was a good reason for their reluctance. I couldn't shake the feeling that this so-called "fresh start" might just be another thing we wouldn't talk about in years to come.

3

WEDDINGS
BEFORE
DEPARTURE

IT DOESN'T TAKE LONG *for me to become captivated by this solitary, lonely soul called Jock.*

I can't stop thinking about the emotions he stirs in me. Meeting him makes me reflect on myself. As I grow to know him better I feel on some deep level that I understand him. He was shunned or unable to connect with others. I know that pain.

He gradually becomes a fixture in my life. I count the days to our visits and during these many months, I feel my personal connection with him grow deeper.

Our trips to the river always take the same routine. Mike, another research assistant called Steve and I launch the boat and putter into Jock's territory. Sometimes we are joined by Mike's other assistants, too. Jock sees us and stops circling his decrepit launch. He races over and greets us.

At first we hear a "puh," the sound of him exhaling and inhaling through his blowhole. On very still days, with little boat traffic in the river, you can hear a dolphin long before seeing one. I still associate "puh" with excitement and exhilaration, knowing a dolphin is nearby.

When he eventually reaches the research boat he blows raspberries and makes clicking sounds. I come to know it as a dolphin version of: "Hello, you're back."

I love those sounds. Out of his blowhole emerge a remarkable range of high- and low-pitched raspberries and clicks, which he delivers in long bursts or short, fragmented spurts. He seems to be trying to communicate with us in a way we simply don't understand but which always brings a smile to our faces.

Every time I greet Jock verbally, he responds with either a short or long reply.

Mike sometimes drops an underwater microphone from the boat to record Jock's chatter. I'm not sure what he learns from those recordings other than Jock's vocabulary seems to be diverse. No matter what it really means, I feel that Jock is simply trying to connect with us on a different level—using sound and vibrations.

Listening to him and feeling his sonar only add to the amazing experience of physically being in the water with him. Every time I take the plunge it is wondrous, new and exhilarating.

On almost every trip I make with Mike over almost three years, we check in on Jock before going farther afield to find other dolphins. I never let on that seeing Jock is the highlight of the research trips for me and the impact he's having. I feel no one, especially a scientist, would really understand. The truth is that I just want to stay and hang out with Jock, although I know we need to scan the river for other dolphins.

ONE PARTICULAR DAY of dolphin searching dawns without even a whisper of wind. The water shines like polished silver. On days like this, dolphins are easier to spot; their dorsal fins rise like half-moons over the horizon.

Seeing newborn dolphins in the wild is rare. I have been lucky to see a few babies. This is one of those rare days. While I knew my friend Jock was a loner, I also knew this wasn't normal. For most dolphins, family and friends seem to mean everything.

It doesn't take long to find our first pod—a group of four or five. They seem to be circling something. I haven't seen this behaviour before. I wonder if they are herding fish into a funnel so they can all take turns to feast.

We approach slowly for a closer look. The dolphins pay no attention to the boat and continue swimming, diving and sur-facing. They are orbiting something.

Mike turns off the motor. "I think we might have a calf."

We hang back, not wanting to interrupt them. I hear Mike's camera. The dolphins are vocalising through their blowholes— burring, clicking and forming loud raspberries that reverberate across the water. They exhale loudly, rapidly, like runners trying to catch their breath.

I recognise some of them. They are all females, who are usually dispersed among different pods. Today, they have come together. In the middle of the group, a small rubbery form emerges. The new calf, perhaps only a few days old, is about 100 centimetres long, which is tiny compared with the adults. Its skin is crinkled, folded like origami. A little dorsal fin sits cock-eyed on its dark-grey wrinkled back.

I have never seen anything so cute. I hear myself giggling.

The calf is trying to use its tail. Its movements are haphazard, a far cry from the smooth movements of the circling adults. Then I make out a large grey form underneath the youngster. An adult is supporting the calf on her snout, balancing it perfectly so its tiny belly-button blowhole just touches the surface.

A faint "puh" is followed by a gurgle as if it has swallowed water. "Is it having trouble breathing?" I ask.

Mike shakes his head. "Don't think so. Mum's trying to teach it to breathe. Dolphins teach their babies, who then teach their own babies. Sometimes, the little ones are so clumsy and unco-ordinated, it takes them a while to get the hang of it."

The baby's mum perseveres, then breaks away to come up for air. One of the other adult females takes her place. They are working together to ensure the survival of this teeny, crumpled wonder.

"Mum will keep going until her baby can breathe fully on its own. It won't take long," says Mike. "She and some of the others could be related. They'll protect it and make sure it has a fighting chance. That calf may well stay with Mum for years. Maybe they'll remain close forever."

An awareness is dawning on me. Although Jock has ended up on his own, by accident or design, this is how things should be. The bond between a mother and baby. A mother putting her family's welfare above her own, ensuring the family's safety, security and welfare at any cost. To me, this is love in its pure and unadulterated form, with the unflinching support of family and friends.

I am watching something that will profoundly affect me. The natural world is allowing me to experience and become immersed in something I had never believed possible. Unqualified devotion. This group of dolphins are single-minded in their mission to ensure this new life, their latest pod member, will thrive and become another integral part of nature's complex, co-dependent web of living things.

It doesn't matter what kind of dolphin this little baby will grow up to be, what it will look like or who it will hang out with. Its mother just wants it to survive.

AFTER THIS EXPERIENCE, I found myself wondering again, how had Jock come to be on his own? What had happened? It saddened me to think I'd never know how he had lost his mother.

I thought about my own mother. I realised that my desire for her to give me this kind of devotion would never be fulfilled.

Unlike my father, she did try to show love towards me. But it was always unpredictable and seemed conditional.

Watching those female dolphins made me question if perhaps she had never been taught how to be a mother. Her own died when she was so young. If she hadn't been nurtured and loved herself, then how could she reciprocate? How would she know how to do it? I knew I would have to accept that and forgive her, if I wanted to heal myself.

I knew sometimes she did try hard, especially when my sisters got married.

FOR WEEKS, A DARK CLOUD had descended on the cottage. Sombre skies, constant rain and threatening storms reflected the gloom inside.

My sisters seemed despondent, their mood constantly dark. They snapped at me often. I frequently heard arguments between them and my parents.

From the fragments I gathered by listening outside closed doors, they both wanted to stay in England and get married to their boyfriends. Neither wanted to go to Australia with us. But they needed my parents' permission. Their boyfriends were a few years older than they were.

Ana's plan was to marry Neil and move away. Dot wanted to marry Pat and stay in the cottage.

I wasn't privy to many more discussions but, somehow, Mum and Dad must have agreed to let them have their way. They would both get married three months later, just after their seventeenth birthdays, and before the rest of us left for Australia.

There was so much to plan. Dot was getting married first and I was one of her bridesmaids. I was excited. The weddings were to be held at a local church just outside Saltash.

The church reminded me of the gingerbread house in "Hansel and Gretel." It was small and made from dark grey blocks of stone, like those jutting out of the ground in Dartmoor. Its faded terracotta roof shingles were dotted with clumps of jade-green moss. Its wooden door was chunky, with heavy black metal trim. I thought it looked like the entrance to a dungeon. But the grey church was set in a delicate English garden, surrounded by roses of many colours.

Both Ana and I wore full-length lemon dresses embellished with little white daisies around the neckline. With long, flowing bell sleeves, the dresses were like ones worn by medieval royalty. We each carried a posy of bright yellow daisies. Dot, Ana and I travelled to the church in a sleek, black hired car. When we arrived, the heady scent of roses made me feel like I was walking through a forest of flowers in one of my fairy tales. Overhead, a powder-blue sky was dotted with marshmallow-like clouds.

My mother had bought a new dress for the occasion. It was fifties style with a fitted bodice and full skirt. She looked beautiful. My father wore his good grey suit, white shirt, skinny black tie and a grimace. My brother also wore a suit. He looked smart.

Dot wore a simple white knee-length dress with bell sleeves like mine. A short veil cascaded from a ring of daisies on her head. Her thick strawberry blonde hair was pulled back into a loose bun at the base of her neck. She seemed to glow as she walked down the aisle with my father, the smile never leaving her face.

The ceremony was attended by a small crowd, mostly mutual friends and Pat's family. Afterwards, there was a modest celebration at the local pub. Dot changed into her honeymoon outfit, a striped, blue mini dress. She left the pub with her new husband Pat in his forest-green Ford Cortina. Beer cans on string were tied to the tow bar, which jangled, clunked and clanged as they drove away. Although it had been a happy occasion, this was a sad sound. It meant she was leaving.

About a month later, another wedding, in the same location. I was bridesmaid again. This time, I wore a lime-green dress adorned with layers of chiffon in tear drops, which ruffled in the warm breeze. It was sleeveless with a high neckline. My hair was fashioned into an elaborate high bun, studded with flowers. I felt pretty.

My father wore the same suit. My mother had a different dress, in the same fifties style.

Once again, the skies were blue, with cushiony clouds. Ana looked stunning. Her dress was a little more elaborate than Dot's—white satin sprinkled with tiny iridescent pearls. A long lacy veil trailed behind her. She seemed to glide down the aisle; I imagined she was wearing roller skates under her dress. After the ceremony, there was another small celebration at another local pub.

After their honeymoons, Ana rented a flat with Neil near Bristol, many miles to the north. Dot moved into Pat's flat in Saltash, with a plan to move into the cottage after we left. She told my father they would take out a loan to buy the cottage. He agreed. I was surprised he seemed so supportive. The cottage had belonged to his parents, who had left it to him, so maybe he was happy it was staying in the family.

With my sisters gone, the house felt so quiet. They popped in to visit occasionally; both seemed happy. My mother and father, surprisingly, were on their best behaviour during these visits. I wondered why they didn't seem more upset at the prospect of life without the twins. I knew I was going to miss them terribly.

Life was busy. There was always something to organise, rummage through, throw out or pack. For a while, my parents fought less, but the arguments and bruises still erupted from time to time. My father continued to ignore me, unless he was barking orders or complaining about how much "junk" I wanted to take.

I watched as my toys were thrown in the garbage but refused to let my stuffed giraffe Jody meet the same fate. Jody was a present from my mum and, along with Puss, was my nightly bedtime companion. I also insisted on keeping a few of my favourite books, including my most-treasured book, *The Lion, the Witch and the Wardrobe*.

During those weeks, my world was slowly being dismantled, with the remnants stuffed inside suitcases and trunks.

Our departure day came too soon. It was winter 1976. It was cold, with rain pelting the roof. It seemed appropriate.

I held Puss, my face nestled in his thick black fur. He smelled like coal and grass. He was purring. I was going to miss him so much. I knew Dot was moving in soon and would look after him. But I worried about him being lonely without me. Who would he snuggle up to at night? Would Dot read to him, as I did?

Mark was slumped on the couch, immersed in his favourite comic *Dennis the Menace*. I wondered where my sisters were. Maybe they'd be at the airport.

My father had sent some of the bigger pieces of luggage ahead but we still had a sizeable pile by the front door. Much of our furniture had been sold or thrown out. The cottage felt empty, hollow, like its soul had been stolen.

My mother and father were buzzing around, bumping into each other, becoming increasingly frantic.

"Have you got the passports?" My father's voice bounced off the walls.

"Yes, I've told you a thousand times, they're in my bag," Mum said.

"Mark, stop reading that pile of rubbish and get here," my father said. "Melody, put Puss outside. I've left the garage open for him. He can eat and sleep in there for a few days."

Reluctantly, I carried Puss to the back door. I didn't want to let him go. I gently put him down and he slunk out. "Bye Puss," I whispered.

I was so upset about leaving him. It didn't seem right to abandon him like this. He'd been my constant companion for so long. He was part of the family, albeit a broken one. Why did it seem so easy for everyone else to leave him there and hope Dot would look after him? My heart felt heavy. I promised myself I would never forget him.

I picked up Jody and held him tightly to my chest. He wasn't warm and alive like Puss, but I felt a little better.

A horn honked. The taxi was here. The shiny, black chunky car reminded me of pictures I'd seen of London cabs, but bigger. It had a long hood with a large silver grille and enormous headlights.

The driver got out, a big-bellied man wearing a grey woollen jumper that I suspected had once been white.

"Off to the airport, are we, aye?"

"Yes," my father replied. "Give us a hand with the baggage."

"Bloody hell, you got enough 'ere? Goin' for a long time are ya, aye?"

I couldn't place his accent; it wasn't Cornish. Welsh maybe.

"That's the plan," my father responded. "Get in," he barked at us. "I'll lock up."

My brother scrambled into the back seat of the cab and I sat next to him. My mother squeezed in next to me.

Mumbling, the driver took his spot. I watched my father lock the front door and put the key under the potted lavender. He got into the taxi. The driver mopped his face.

"Heathrow," my father said, pushing back his damp hair, which had started to curl.

We pulled away from the curb. My stomach was churning. It would be a long drive.

I turned to look out the rear window. Raindrops distorted my view. The cottage became blurry, taking on a dream-like quality. I squinted, wishing I could wipe the window. Before long, the whitewashed walls disappeared behind the crest of the hill.

Time passed. Silence, just the thrum of the engine, the clunk of gear changes. The driver started to whistle.

"Are Dot and Ana going to be at the airport?" I asked.

"No, they can't. They're working," my father replied.

I could feel tears pooling in my eyes. "But I never got to say goodbye properly."

"You'll see them again," said Mum, handing me a hanky. "They'll come to visit."

Tears trickled down my cheeks. "I don't want to go." My lips trembled. "Can't I stay here?"

The driver looked into his rear-view mirror. He made eye contact with me, then turned his attention back to the road.

"They've got their own lives now," Mum said. "You can write to them."

My father said nothing. He stared ahead.

Mark was looking out the window. I grabbed his hand.

He glanced at me, and squeezed back. It was just him and me from now on. My heart was pounding in my chest.

A place couldn't change people, could it? I thought. We'd be on the other side of the world, where I couldn't join my sisters on country drives. They wouldn't be there to hug and reassure me. Their hands would no longer cover my ears, muffling my parents' screams. I was scared.

The drive seemed endless. Little was said, apart from my mother asking if we wanted a butterscotch or if we needed to stop for a pee. After an eternity, we arrived at the airport. Cars and taxis were scooting in every direction, with a cacophony of horns. I'd never seen so many cars and trucks of different makes and sizes.

"That's a Maxi!" Mark's voice took me by surprise. "I haven't seen one of those before."

His face was animated. He was excited. He had a collection of Matchbox cars, some of which he'd smuggled into his luggage.

I smiled at him. My tears had dried on my face.

"Righto, international departures just over there," said the driver as he pulled up.

My father fumbled for his wallet and counted out cash into the driver's hand. They got out and started taking the cases from the rear.

"Let's go," my mother said.

We clambered out and stood on the curb. I couldn't believe there were so many people. Smart suits, colourful scarves, impossibly high wedged shoes. I noticed a black woman dressed in a bright orange pantsuit with flared cuffs. A mass of tight curls was piled on her head and secured with an indigo scarf. The beautiful contrast of colours struck me. Ebony skin, vivid orange. She looked so stylish.

Overhead, there was a roar of engines. The planes were so close, I felt I could touch them. How could something that big fly, I wondered.

My father had collected two carts for our luggage. We navigated the throngs of people to the ticket counter. My parents showed their passports and unloaded our bags. Apparently, we were stopping in Asia, at a place called Singapore. The woman at the counter put tickets on the cases, which disappeared through a hole in the wall.

Then we walked for what seemed like miles and stood in lines with no obvious end. The airport was like a city, with people and bags everywhere. Smart-looking pilots walked next to glamorous women, who I assumed were air hostesses. Everyone was in motion.

Finally, we boarded the plane. It was my first time. I sat next to Mum, who buckled me in. Now, I realised there was no going back. I started sobbing, wanting to go home.

"For God's sake, Melody. Just stop it, you're making a scene," my mother snapped.

The engines fired up with a rumble.

"Look out the window," she said.

I did. It was spotted with raindrops.

We moved slowly and then, as the engines roared, I was thrust into the back of my seat. I craned my head to look out the window. Through the water droplets, the world was falling away. The airport looked like a Lego model, then a white shroud enveloped my view. Who knew you could fly through clouds? Once we were above them, it looked like someone had spread out a white feather eiderdown beneath us. I wanted to jump out of the plane, roll up in it and sleep.

I felt my mother's hand on mine. "It'll be alright," she said.

I doubted it.

I WOKE WITH A START, my heart racing. I'd been dreaming about black spiders scurrying up cherry trees. I stood and peered at Mark, sitting with my father across the aisle. Mark was reading a comic. My father was dozing. Beside me, my mother, wearing headphones, was flipping through a magazine.

Hours passed. I entered the wardrobe filled with furs and returned to Narnia.

We eventually landed in Singapore.

The cabin sprang to life. Passengers were on their feet, grunting and muttering loudly, pulling bags and packages down from lockers and out from under seats. Above the excited chatter, a man's voice boomed from the speakers. He welcomed us, and explained the weather and time.

My mother stood, stretched and yawned. She reached down, grabbed her handbag from under the seat and turned to look at me.

"Better get up. Make sure you don't leave anything behind."

I rummaged through the seat pocket in front of me, finding nothing. I clutched Jody tightly, my book feeling bulky in my coat pocket. My mother and I nudged into the queue of people standing in the aisle. We shuffled forward and stepped off into a tunnel.

It felt strange, not solid, but hollow like an enclosed bridge. The floor clunked as I walked, my calf muscles complaining as we trudged up the slight incline.

Ahead, our hostesses were smiling, standing at the entrance to the terminal. They were wishing everyone "Good day," as we walked past.

We entered the terminal building. I felt clammy. I took a deep breath. The air was warm, moist, cloying—such a contrast from the stale, frigid air on the plane. I took off my coat.

I looked around, disappointed. I expected it to look more exotic, oriental. I wanted to see Asian women wearing intricately embroidered dresses, but everyone was dressed in normal clothes.

A huge black pot was overflowing with the most unusual flowers I'd ever seen. Their delicate pink petals looked like they'd been intricately carved from icing, like the little yellow roses on Dot's wedding cake. Their petals surrounded vast, gaping mouths with spotted tongues poking out at me. They seemed to be making fun of my arrival.

"What kind of flowers are they?" I asked my mother, who was staring at the exit, tapping her foot.

She turned and studied the flowerpot.

"I think they're orchids. Singapore is well known for them. They grow everywhere."

She turned back to the tunnel. I couldn't take my eyes off the exquisite flowers.

"About time," I heard her snap. My father was holding my brother's upper arm. It struck me again, he never held our hands.

"Your son needed to go to the loo," my father said.

Mark didn't raise his eyes from the floor.

We walked in silence, then waited in a long line for uniformed men to stamp our passports. My father was getting agitated. Sighing and muttering, he declared loudly that we needed to be at the harbour by 5 o'clock and "these people" needed to hurry up.

An hour or so later, trollies laden with luggage, we went to find a taxi.

The stuffiness inside the terminal was nothing compared to the suffocating mugginess outside. Steam seemed to rise from the footpath. My pants were sticking to my legs. I sucked in a deep breath. The air smelled heavy and rank. It was unfamiliar—sweet, damp, musky, smoky, with a slight scent of cloves. It vaguely reminded me of Christmas desserts and rotting vegetation, both pleasant and unpleasant at the same time.

"Bloody hell, it's humid," my mother said. She twirled her limp curls around her finger.

We got into a bright yellow taxi. Our driver wore a blue turban. He helped my father load our luggage in the trunk. On the route from the airport, I became mesmerised by the plants along the roadside. Towering palms and trees with enormous flat, shiny plastic-looking leaves. Thick lush shrubs vied for space. Many were laden, and some overloaded, with brilliant pink and red flowers. They too looked unreal, as if skilfully crafted by hand. The colours were all so intense, so vivid, it was like I was seeing clearly for the first time.

We passed lots of people riding bikes, some of them towing rickshaws. The rickshaw drivers all looked ancient. Their wiry, spindly muscular legs were bulging, bare feet pushing down on the pedals, the lines on their nut-brown faces deepening with exertion. Some had cigarette nubs hanging from their mouths.

Trucks spewing acrid smoke rattled along the road, their open trays filled with dark-skinned Asian men. They huddled together, packed tightly. I could see the whites of their eyes, their faces seemingly void of expression as they passed us.

The buildings were a hodgepodge of old and new. Small, brightly coloured terrace homes were wedged between grey concrete skyscrapers, like they were being bullied and pushed out of their tiny spaces. I felt sorry for them.

4

THE JOURNEY
TO ADELAIDE

EVEN WHEN THE *inflatable is upgraded to a half-cabin fishing boat a few months after I first meet Jock, he still recognises us. On our first visit in the new boat he hesitantly approaches the hull and turns on his side, one eye surveying the new craft and its occupants, perhaps to make sure he's approached the right people.*

Every time I interact with Jock it is special. I never forget how privileged I am to be in his company. He often follows us into the quiet, mangrove-lined backwaters, away from the prying eyes of other boaters and fishers. I watch, transfixed, from the stern as Jock's mangled dorsal fin cuts through the water behind us. His silvery body glides effortlessly, keeping a safe distance from the motor.

Then, finally, comes the excitement of jumping in to play. It seems to me that Jock feels the same. He sidles up to the boat and nudges the ladder. Sometimes he gets impatient and bangs his snout on it as if telling us to hurry up.

The aluminum ladder is located on the back of the boat. The top half is secured to the boat with a latch, allowing it to be stored above the water line safely while we are moving. On many occasions Jock bangs the ladder so hard, he manages to unhook it and the bottom half unfolds out into the water. Once the entire ladder is extended, he hangs in the water next to it, blowhole just above the surface and strikes it rapidly, perhaps the dolphin version of impatiently tapping feet.

He seems to know that the ladder is the way for his human companions to get in and out of his environment. Once I am in the water with him, he leaves the ladder alone.

During my early interactions with Jock, I wear flippers and he seems to be fascinated with them. I swim, kicking my legs as fast as I can, and he follows in my wake, his snout millimetres from the frenzied flicking of my fake flukes. He seems to enjoy the feeling of the water being displaced by them. When I stop he runs his body alongside the fins, scratching himself gently on the firm rubber. He seems to enjoy the feeling of the material and is focused on finding as many ways as possible to explore my flippers with whatever part of his body he can rub them against.

On days when I'm not wearing flippers, he makes do with my feet, but I'm sure he wonders where my interesting, artificial fins have disappeared to.

When I get back onto the boat, he turns his attention again to the metal bridge between our two worlds—and resumes his incessant tapping. This seems to be his strategy to entice me, Steve or Mike into the water once more.

He often rests his snout on one of the lower rungs while watching us dry off on the boat. Despite his protestations the time always comes for us to secure the ladder and head off to other parts of the river so that Mike can document the behaviour of other dolphins.

Although we spend an hour or so with Jock, it's still early and we have a morning of dolphin spotting ahead. It's one of those days that can't work out what it is. One moment, weak strands of sunlight shoot bronze arrows across the water. Then, in an instant, they vanish without a trace, disappearing as quickly as a magician's trick. Nature's erratic alchemy in action.

Like the sunshine, the wind is indecisive. Powerful gusts are followed by long lulls, like the initial blasts have taken too much energy, leaving behind a lethargy. In turn, the river waffles, churning wavelets that wane soon after they form.

Seated on a plastic box in the back of the boat, I pull up the collar of my jacket against the wind. Mike has bought a new pair of binoculars and I'm trying them out. As I scan the water for fins, I fiddle with the focus. About 50 metres ahead, I notice flicks of water along a stretch of bank. From my vantage, it looks like small waves crashing on shore. I know that, despite its sabre-rattling, the wind today doesn't have the power to whip up waves like that. Something has to be splashing around up there. We head over to investigate.

The water is too shallow along the bank for us to approach closely. We don't want to end up stuck on a sandbank with the receding tide. Mike throws the boat into neutral, only clicking it into gear when we start drifting too close. We don't want to interrupt the spectacle.

A string of small golden rockets is catapulting, darting along the water's edge. Flashes of iridescence. Mullet and mackerel are leaping for their lives.

A flurry of fins zip back and forth, a frenzy of tails thudding the surface, herding the fish and forcing them to flee ashore. Stunned and confused, the dolphins' prey lunge into the shallows, allowing their pursuers easy access.

Initially, the action seems helter-skelter but, as we watch, a strategic pattern emerges. This is a calculated, collaborative attack. Like an elite squadron, the dolphins work together with precision. A couple of dolphins will scare the fish, rounding them up along the bank. Others will take it in turns to sweep through the treacherously shallow water, devouring fish as they swim. After feasting, they swap roles.

Occasionally, one dolphin will pop its head out of the water, like a periscope. It quickly eyeballs the scene before returning to herding. Mike calls this "spy hopping"—seeing what is going on above the surface.

One of the dolphins breaks off and speeds towards the boat. I race to the bow to get a better look. It's in hot pursuit of a long, skinny snook just below the surface. The snook darts sharply, the dolphin following with its snout just centimetres behind. The snook scoots and dashes around the boat but, no matter how erratic its movements, it can't shake its pursuer. I wonder if the dolphin is having fun, playing with its food before eating it. The drama is transfixing, like watching a high-speed police chase unfold on the news. In this case, however, there is little hope of the fish evading capture. The pair dive. I assume lunch is now being served.

The dolphins suddenly break away. The feeding frenzy is over. The energy of the moment dissipates almost as quickly as it began. The pod reforms and meanders calmly back into deeper water. We follow them, keeping our distance. They double back, approach the boat and start to weave in and out of our bow wave.

As I watch them roll and entwine in the waves, a familiar feeling of elation sweeps through me. I've begun to feel this sense of joy more and more as I witness dolphins in action in the wild.

At the time, I'm only beginning to realise what a crucial role they would play—Jock in particular—in helping me heal from the pain of my childhood. Watching them also reminds me of my first dolphin encounter, when I was on board the ship from Singapore to Australia.

I GASPED IN WONDER at the port. Huge ships were tied to the dock by bleached ropes, thicker than my fist. Giant yellow cranes lined the water's edge, lifting metal containers from the ships high into the sky, like they weighed nothing. A tinny, industrial smell mingled with the salty air. I heard people shouting in a language I didn't understand. Other big ships were moored off shore—dozens of them, sitting still and silent. Some of their bulbous hulls looked battered, rusted. I wondered how long they'd been at sea.

We walked to the passenger terminal. Looming ahead was our passage to Australia, with the letters "CTC" painted on its side. The paint looked old, part of the "C" was missing.

We boarded. Our sparse cabin had four bunk beds. I jumped onto one and saw cockroaches scamper up the wall.

My mother shrieked. "Cockroaches, they're bloody everywhere!" She looked accusingly at my father, who was sitting on the opposite bunk. "You cheap-arse bastard. What kind of bloody ship is this?"

"A Russian one," he said, deadpan.

"It's filthy. God knows what bloody germs we'll catch."

"Will you just shut up. It's only a week or so. God, you do nothing but whinge."

Mum threw her bag onto a top bunk, then followed it up, slumped against it and stared into space.

With my parents bickering daily, Mark and I often fled the oppression of the windowless cabin.

The deck sometimes crunched under our feet. Patches of dark green paint were peeling away, revealing dull metal. Faded steel safety railings with weathered wooden planks enclosed the space. Hugging the cabin side of the deck were rows of hard timber deck chairs with faded blue covers. We lounged on them. Mark read his comics and I returned to Narnia.

We spent hours outside, our elbows on the railing, look-ing out to sea. It seemed to go on forever with no land in sight. I asked Mark how the world could be round if the ocean was so flat. He didn't know. We argued about whether the water was blue or green or purple. I loved hearing the rhythmic slapping of waves against the metal hull. It soothed me like a lullaby.

Some days, the sea was calm and quiet. Sunlight glittered off ripples, turning them glossy. I wanted to touch the water's surface, thinking it would feel like velvet. Other days, it was choppy, the familiar salty smell becoming more intense. White-capped waves sloshed at the ship's side. The silvery froth was hypnotising; I marvelled at how the sea could change so quickly. It switched as rapidly as my mother's moods. Some days, the ship rolled from side to side. We saw lots of people run to the railing and vomit.

Then one day, Mark and I were out on deck, watching the mottled shades of the ocean, dreaming about mermaids and pirates when, out of the blue, a group of fins appeared.

I was so excited, I wanted to run inside and tell all the passengers we were being escorted by real, live dolphins. But I didn't move. I was spellbound. I couldn't believe how effortlessly they kept up with the ship—they didn't even seem to be trying. Although I was a good swimmer, having learned at an indoor pool in Cornwall, I knew I could never keep up with these creatures.

The pod stayed with us, diving, surfacing and leaping close to the ship, playing in the waves the ship created. Mark and I stepped up onto the bottom rail to get a better view, although we knew we shouldn't climb all the way. As much as I loved watching the dolphins, I didn't want to end up overboard with them.

Mark said he'd heard that dolphins attack sharks by ramming them in the gills. He also told me that, in the old days, dolphins saved lost sailors at sea. They even carried shipwrecked survivors on their backs to safety. I was loving these slinky animals more by the minute.

That evening, I dreamed of enchanted undersea cities, in which dolphins played games with children. Circling us were sharks with wide, snapping jaws filled with huge, jagged teeth. But I wasn't afraid. I trusted the dolphins to keep me safe.

The sea and the regular sight of dolphins sparked our imaginations. I wondered, could I ride a dolphin instead of a horse? Mark and I told each other ever-more fantastic stories.

Mark talked about underwater cities ruled by scaly, bearded half-men, half-fish kings, who were terrorised by monsters of the deep. I imagined growing a tail and turning into a mermaid princess. Would I have to kiss a fish to turn into one? Mark became a submarine captain, who discovered an undersea metropolis littered with bags of gold snatched from sunken pirate ships.

Sometimes, our story-telling was interrupted by deep rumbling blares, which seemed to bounce off the water from faraway cargo ships. I wondered how those huge bulky vessels floated, with so many crates stacked on top of them.

When we ventured indoors, I hoped we would bump into one of the waitresses. They always said "hello" and smiled brightly when they saw us. They bent down and, in broken English, asked us what fun things we did that day. We told them excitedly about seeing dolphins and our wondrous, hidden world under the sea.

The waitresses were all young, about the same age as my sisters. Most were dark-haired, tall and slim with creamy skin. They wore uniforms with a dark blue ruffled apron, white shirt, navy skirt and always wore bright red lipstick. They moved slowly, gracefully and purposefully, like ballerinas. I wondered, why did they want to work on a ship? They could be models, I thought, like the ones I'd seen in my mother's magazines.

When the captain announced we were almost at Sydney, I was sad. As we were disembarking, our waitresses stood in line at the door, smiling and shaking hands. Two of them leaned down and kissed Mark, leaving bright red splotches on his cheeks. Beaming, he turned as red as a tomato. I will never forget that smile; it was so real.

After we disembarked, we collected our luggage. We'd arrived in Sydney Harbour.

The first thing I noticed was the water. It was deep turquoise, stretching for what seemed like miles. Sunshine glinted off it, as if Tinker Bell had touched it with her wand. Gulls cried overhead. The harbour teemed with vessels of all shapes and sizes. Ferries packed with passengers zipped here and there, leaving white frothy trails behind them. Smaller boats with "Taxi" written across their canopies zig-zagged between the ferries. I wondered if dolphins lived here and if they ever collided with all the boats.

I could hear the excited chatter of people on the dock. There seemed to be hundreds of them, some with luggage, while others were casually strolling, stopping occasionally to look at the bustle on the water. I was mesmerised by all the activity.

"Oh my God," my brother said.

I followed his gaze. A black metal arch erupted from the water. I could see cars crossing it. It was much bigger and more intricate than the bridge across the River Tamar back home. I stared in awe.

"That's the Harbour Bridge," my brother said. "It's a wonder of the world."

"Wow." I studied the convoluted lace of black metal. "It's huge. It looks like someone crocheted it."

"Honestly, Melody. You're such a stupid girl." My father jumped in. "It's the world's biggest steel arch bridge. Most of the steel came from England. It's a bloody engineering feat. Millions of rivets are holding that thing together." He turned to my brother. "But it's not one of the seven wonders of the world." He rolled his eyes and hailed a taxi.

We travelled up and down steep hills. We passed some buildings that reminded me of home, except the stones were darker. They looked sooty like our old fireplaces. Buses, cars and people were moving everywhere. Maybe I was wrong about the desert. I gasped at the huge structures shooting up from the city skyline. They seemed to touch the flimsy gauze of clouds overhead. I wondered, did people on the top floor ever look down at clouds, like I did from the plane? Where was Blacktown, was it near the city?

I glanced at my mother. She had sallow bags under her eyes, her hair had lost its curl, and her expression was surly. I decided not to ask.

SYDNEY AIRPORT LOOKED like Singapore Airport, but without the orchids. It had the same white-tiled floor and soullessness. It seemed massive. Shops displayed toy kangaroos, wearing green hats with tiny corks dangling from string. The word "Australia" was everywhere. Chocolates from Australia, nuts from Australia, wine from Australia... they must make a lot of stuff here, I thought.

Eventually, we boarded another plane, much smaller than the one we took to Singapore.

I sank into my seat and found my place in my book.

A while later, the captain announced we were approaching Adelaide. Out of the window, I could see beige fields. It looked dusty. Clusters of withered trees dotted the flat landscape, which reminded me of the parched plains in cowboy movies. I wondered if there were tumbleweeds. As if from nowhere, mounds started to appear. Although they broke the monotony, they too looked desolate and drab.

Houses came into view. They seemed lonely, sitting in their vast, empty fields. I thought about the rich, emerald hills of the English countryside and felt a pang. This place seemed colourless and lifeless. My bottom lip started to tremble. I sat back in my seat. Dear God, I would be living in the desert with kangaroos and spiders, after all.

With a bump, the wheels touched the ground. The plane roared as it slowed. We had arrived in Adelaide. My mind spun and my stomach knotted. Through the window, the airport seemed small and deserted, unlike Singapore's colour or Sydney's vastness.

My home seemed a million miles away. I had landed on another planet, not in another country.

The passengers stood all at once, grabbing bags from overhead lockers. We shuffled out. I stepped onto the metal stairs leading to the tarmac, and into a furnace. The sun was blazing on my face. I squinted, feeling like I'd left a cave and emerged into blinding light.

"Oh dear God, it's stinking hot," said my mother.

The stairs felt wonky as we clambered down. The handrail felt like the inside of an oven; I couldn't touch it.

It felt even hotter on the tarmac. We waited for my father and brother. Sweat trailed down my back. Mark looked feeble as he descended. He was thirteen but his twiggy pale legs, jutting out from his long shorts, made him look like a young bird—wobbly, trying to find its footing.

AN HOUR LATER, we arrived at the immigrant hostel. It was a white, featureless concrete block, which sat in the middle of a dark grey parking lot. The asphalt looked feverish. The heat here was so different from that in Singapore. It was dry, harsh and biting. Inside the building, it was slightly cooler.

"Hi, welcome!" A short, plump woman with almond skin and spiky brown hair walked up to us, her arms open. She had the strongest Australian accent I'd heard on my journey so far. She was wearing a plastic badge that said "Sharon."

"Good afternoon," my father said. "I'm Neil Horrill. We're staying here for a couple of weeks."

The woman smiled. "Fabulous. Just let me check you in and get your passport details." She turned to my mother. "You can call me Shazza, dear. Let me know if you need anything, or if you just want a chat. Looks liked you've got your hands full." She winked at Mark and me.

"I'm sure we'll be fine, thank you." Oh no, my mother was using her posh voice. "I really don't want to talk. I'm tired and I just want to get to the room, actually." Her tone was curt.

Both Mark and I looked at the floor. I noticed the mission-brown carpet wasn't carpet at all. It was tiles that looked like carpet. I could see the joins.

Sharon's smile disappeared. "Oh, okay then. I'll sign you in and show you to your room."

My father went to a counter with Sharon and filled in some paperwork. He showed her the passports. She jotted down the details. We followed her down a narrow hallway. It was beige, its walls unadorned. Unremarkable. Doors led off it. She opened one.

"Here ya go," Sharon smiled. The smile looked a bit forced. Like the ship, the room had two bunk beds, but it was bigger.

"Well, this'll just have to do, won't it?" My mother snorted as she looked around. "At least there aren't any cockroaches."

"For God's sake, you'll have your bloody house soon. The money's been wired. But you'll probably whinge about that, too." My father scowled.

"I haven't even seen it," she retorted. "You'd think I would've bloody seen it before you bought it."

Sharon said quickly, "Well, if you need anything..."

My parents ignored her. She hurriedly left.

I couldn't hold back any longer. "Mum!" I cried. "My bum and legs are really itchy. It's getting worse."

My mother and father stopped glaring at each other and looked at me.

"They were itchy in Sydney, but the itching's got worse," I said, scratching my buttocks to prove my point.

My mother sighed. "I saw a bathroom down the hall. Come with me." She grabbed my arm roughly.

Once in a cubicle, she pulled down my pants. "Oh Christ," she sighed. "Wait here. I need to get the nurse."

I stood with my pants around my ankles, worried I'd contracted measles or chickenpox or maybe some rare Russian disease. I craned my neck, trying to see my bottom.

Mum returned with a woman wearing rubber gloves, who said she was the first-aid officer. She kneeled down and examined my rear, poking at it. "Looks like impetigo," she said. "Probably picked it up on the way. Did you stay anywhere unsanitary?"

"A cockroach-infested ship," my mother said.

"Well, it's contagious. I'll get you some cream. If it doesn't go away in a few days, you'd better get to a doctor. Your daughter will have to isolate in the room while these sores are open." She looked up at me. "Sorry, love."

My mother thanked the nurse. We walked back to our room. As soon as she walked in, my mother yelled, "You stupid bastard. She's got a bloody skin disease from that godforsaken ship you put us on."

My father looked at her, then at me and stormed out. The door slammed so hard, I thought it would fall off its hinges.

I climbed onto my bunk, drew my knees to my chest and nestled my head in my hands. Even with my father gone, the room felt charged.

Clunk, clunk. I looked up as bits of clothing and toiletries flew across the room. My mother was on her knees, throwing things out of her suitcase. She was muttering to herself. Mark was curled up in a ball on his bunk. I put my head back in my hands and closed my eyes. I wished the brave lion Aslan would jump out of my book and save me, as he had saved the children from the White Witch of Narnia.

For the next week or so, I stayed in my windowless room, with only my precious books for company. It was warm and stuffy. I became familiar with the daily noises. The clank of the cleaner's trolley as it stopped outside the bathroom. Sharon's shrill morning greetings. A man singing, his deep voice resonating through the corridors at about 3 o'clock. Children running in the hallway, laughing and squealing. I wanted to go out and play. This was so unfair. My father accused me of sulking. And I was.

My mother slathered lotion onto my legs and bottom twice a day. Eventually, the itching stopped and scabs started to flake off. I was allowed out.

I raced to the playground I'd glimpsed when we arrived. I jumped on a swing, tilting my head back. The sun wasn't

as fierce as on that first day, the heat not as intense. It felt
wonderful to be outside. Closing my eyes, I heard the distant
sound of cars and nearby children giggling. I opened my eyes.
Kids of all ages were running here and there, squealing as
they glided down the metal slide. They seemed to be from
different parts of the world. Some had the same inky skin as
the woman I'd admired in the orange pantsuit at Heathrow.

A girl about my age, with blonde pigtails, thumped down
on the swing next to me. Her face looked like the moon,
perfectly round, but spotted with freckles instead of craters.
Her wide amber eyes stared at me. I felt nervous.

"Hi," she said, her smile revealing rows of metal strapped
to her teeth. "I'm Angie and I'm from Czechoslovakia." Her
speech was stilted and she sputtered as she said where she'd
come from.

She started swinging slowly, keeping her eyes on me.

"My name's Melody and I'm from Cornwall, England." I
stuttered, blushing.

"You have a strange accent," she said. "Really weird."

"So do you," I mumbled.

"Are you shy?" she asked. "Don't be shy, I know my braces
look ugly but I'm a good girl. My mum thinks so, anyway."

I felt a pang. I imagined her parents loved her, telling her
she was a good girl. No doubt they held her hand, cuddled
her and they all hugged each other. Her parents didn't fight;
they didn't hate each other like mine did. She probably had
brothers and sisters she played with.

I could feel tears starting. Oh no. I didn't want to look
like a cry-baby.

She stopped swinging. "What's wrong? Why are you crying?" Her smile had gone. Her eyes widened, staring into
mine. I looked away.

"Nothing. It's just nice to be outside," I lied. "You know how grown-ups say you sometimes cry 'cause you're happy? I'm just happy you came to talk to me." I looked up and smiled feebly.

She grinned. "Maybe you can tell me about England. We should meet here every day and swap stories."

I smiled. "That sounds like fun." But I knew we only had a few days left. "I'd like to know about Czechoslovakia."

She giggled and pulled something out of her cardigan pocket. A foil package. She unwrapped it delicately and took out an oval-shaped cookie with an elaborate scalloped design.

"You tried Yo-Yos?" she asked.

"No." I was intrigued.

"Here." She passed the cookie to me. "Try it. They're good and Australian."

I grabbed it, examined it. It was deep yellow-gold and so pretty, too pretty to eat. I sniffed it, smelling honey.

I bit into one of the most delicious things I'd ever tasted. Sweet, crunchy, I'd never had anything like it. It was nothing like the cookies back home.

"Oh, wow!" I exclaimed, spitting crumbs at my new friend.

"Yes, the Aussie kids put butter on them. I've tried with butter, it's very good," she smiled.

I shovelled the rest of the Yo-Yo into my mouth. I inhaled it. I smiled at Angie. I felt a bit better about Australia. If the worst came to worst, I'd made a friend who I could share yummy cookies with, even for a few days.

5

NO HOME
SWEET HOME

PERHAPS ONE OF THE REASONS *Jock helps me start to heal from my family traumas is that, like all dolphins, he seems to enjoy playing and having fun. As I float face-down on the surface, trying to make out shapes in the murky depths, Jock suddenly appears, sweeping past me, surprising me with the abruptness of his arrival.*

I never know where he is going to surface. He does this repeatedly, each time ensuring with precision that his body skims along the length of my mine as he emerges from below and swims past. He is the epitome of speed, grace and agility, perfectly adapted to his environment. In comparison, I am an awkward, bumbling creature not remotely suited to any degree of speedy, elegant manoeuvres in this watery world.

I imagine him chuckling to himself, amused at the vain attempts of this unwieldy human to keep up with his whereabouts. It's like a game of hide and seek, but the seeker is clueless, and the hider is in complete control, using the element of surprise to keep me on my toes. It seems he knows full well I will never succeed in finding him before he reveals himself.

Jock seems to tire of his game of "guess where I'm going to pop up" after about half an hour and changes his behaviour. At the last point he emerged, a few metres away, he begins very slowly and deliberately gliding towards me on the surface, barely making a ripple, his sonar clicking and burring loudly as he approaches. I feel the familiar subtle jinglejangle of my insides as the sound waves bounce off my body.

Sometimes his games are more vigorous. Fellow research assistant Steve will jump into the water with me and bring the paddle from the boat. We know from previous experiences that Jock loves the paddle. It seems to be his favourite plaything, even without humans attached to it.

Whenever we have the paddle, Jock seems to get even more excited. He bolts towards whoever is holding it, grabs it with his mouth and steals it away, flipping it high into the air with his tail flukes as he swims off with it. Often, Steve and I pursue him, knowing we have no hope of catching up. Tag with a dolphin.

Eventually Jock will let the paddle go. Steve and I will then grab it and hold it at either end above the water. Jock throws himself at it, trying to knock it down. He then circles us, turning on his side and eyeing us, as if assessing his options and devising a strategy. Then he swims away, turns and zooms towards one of us like a torpedo. More often than not it is Steve he targets. I often see Steve's eyes widen as that deformed dorsal fin speeds towards him.

"Oomph." Jock throws himself on Steve's arm and dislodges the paddle. He always gets his paddle. In the background, the soft whirr of Mike's video recorder captures the moment from the boat.

Having fun is one thing Jock has in common with the dolphins who live in other parts of the river. After playing with him I meet another special dolphin who exhibits exuberant, uninhibited playfulness.

Her name is Billie.

SHE SURGES OUT of the water backwards. Sleek muscles straining, she throws herself upright, balancing on her tail. Her body is vertical in the air, out of its element, feeling the Earth's gravitational pull. In a manoeuvre sure to impress Olympic gymnasts, she wobbles her tail rapidly, back and forth, like a samba dancer. Using her powerful flukes, she propels herself backwards across the surface, leaving behind a jagged wake.

Her extraordinary routine ends in a reverse belly flop, her back smacking the water loudly as she lands. Then she disappears, returning to her weightless home. This exhilarating display of strength and agility is how I first meet Billie.

"Wow, that was amazing!" I say, wiping the splashes of water from my face.

"Yep, that's Billie," Mike replies. "She was tail-walking. A little thing she picked up in captivity."

Behind Billie's transformation into the performance artist so loved by many locals around the port is a sad history. As Mike explained, she had first gained fame after being photographed swimming with a racehorse and dog behind a trainer's boat. But apart from racehorses and dogs, at first—like Jock—she had been shy of people and avoided busy areas.

Then one day, an armada of ships came up the river to celebrate Australia's bicentenary in 1988. She disappeared. The fear was she'd fled or been hit by a ship. She turned up, emaciated, in a polluted lake called the Patawalonga, which was used to divert stormwater from urban areas into Gulf St. Vincent. She was captured and taken to a pool in a decrepit dolphin theme park called Marineland, where she was branded with the number 3 on her dorsal. She stayed there until the park closed due to public pressure. She was then released. During her stint in captivity, she was taught to perform tricks for food. One of them was tail-walking.

This was how Billie had become a performer. Despite her chequered story, I admit I had many mixed feelings about Billie. I loved catching up with her. First, she was easy to identify. No other dolphin had a number burned into its flesh. Second, she was a delight to watch. She often performed her tail-walking routine in front of wide-eyed onlookers at the port's wharf. Like all dolphins, she seemed to do it just for the pleasure. It seemed to serve no purpose, though I sometimes wondered if she did it when she was hungry, in the hope that someone would throw her a fish.

While I thrilled at Billie's acrobatic skills, I also felt sorry for her. I had visited Marineland at school and found it a soulless place. Even then, I thought it seemed cruel to keep such magnificent mammals in concrete tanks—devoid of nature, sunlight and freedom—and force them to live in close proximity to one another. Mike talked about the inhumanity of keeping sonic, intelligent creatures in tanks. I loathed the idea of keeping dolphins in captivity, forced to perform for food. To me, it seemed their smile was a curse. People assumed they were happy, even if they were miserable.

Away from the wharf, Billie would often approach our research boat, turning on her side and eyeballing us. Then she'd swim in the wake, leaping high out of the water. I found it astonishing she could trust us, after being in Alcatraz.

What did she make of humans, I wondered? Did she see us as saviours or captors or both? I wondered, too, if she had been scarred from her imprisonment.

It occurred to me, however, that even though Billie was free, she was still displaying behaviours she'd learned in captivity. Although her environment had changed, she still hadn't forgotten her past conditioning.

I had once hoped that moving to Australia, starting a new life in a new house, might repair my family, make us happy. But I feared that, despite the new environment, the old performances between my parents would continue. That the fresh start wouldn't result in any real change.

Watching Billie made me realise that, despite our best efforts and hopes, and our desire to change, it's hard to modify our learned responses to life's challenges. Like Billie, I would never forget where I came from and what I'd learned from my past, but I could learn to embrace them and the scars they left.

This amazing dolphin, who brightened people's lives with her acrobatic displays, inspired me. I could share my learnings and history with others, and maybe turn my painful past into something positive. While I could never fully break free from the trauma of my childhood, like Billie, I could choose to live with it and hope that I could become strong enough to one day use it to help and inspire others.

IT WAS SUMMER 1976. The day before we left for our new house. The heat seemed relentless. I met Angie at the swings. I didn't know then, but it would be the last time I ever saw her. She brought more Yo-Yo cookies. I asked her to tell me about where she'd grown up. I noticed she hugged herself as she told me stories about uniformed men with guns and tanks, lumbering through quiet cobblestone streets.

As she explained how her family always felt scared and dared not speak out against the government, her strange, lilting accent seemed to thicken. She smiled when she talked about how pretty her town was. It had buildings with turrets, which sounded just like the castles in my fairy tales.

She loved her family, she said. Her father, a carpenter, was looking for work in Australia. Her mother wanted to buy land and grow potatoes. She had a baby sister. I told her about Mark and my sisters.

ON THE MORNING of our departure, my father drove an old Nissan van to the centre's front door. He had bought it second-hand. It was pastel blue, with two doors at the front and one at the back that lifted up. Blotches of rust blemished the wheel arches and window frames. It looked dreary and neglected.

The day was searingly hot again. Rivulets of sweat dripped down my neck. The stench of tar rising from the parking lot turned my stomach. I was tired, after tossing and turning in the night, knowing we'd be going to the new house in the morning. Where was it? Did it have a big garden? Were there nooks? Was it near the sea?

As we waited, I nibbled at my nails. My brother was standing next to me. He couldn't stay still and kept shifting his weight from one foot to the other.

"Melody," my mother snapped, as she slapped away my hand. "How many times do I have to tell you, don't bite your nails. It looks ugly and you'll get worms."

I turned my attention to Jody and started pulling bits of lint off his yellow fur.

"Mark, do you need to go to the washroom?" She sounded exasperated.

"No," he said. He fidgeted with the hem of his "I love Sydney" T-shirt, which Mum had bought at the airport.

My father finished loading the luggage. "Right, let's go," he said.

We passed squat, rectangular houses sitting on sprawling patches of sunburnt grass. Many had big trees in their front gardens. Some had silvery trunks, which looked like they'd been splashed with liquid copper. Their leaves were a peculiar colour—muted sage-green—nothing like the deep green of the trees back home. I kept an eye out for kangaroos. I didn't see any.

It all looked so strange. Even the roads looked weird. They were wide and lined with tall concrete poles, carrying thick electricity wires. Later, I learned these are called Stobie poles; they're made of concrete and steel because, here, wooden poles get eaten by termites.

We descended a hill. The van slowed.

"This is it," my father said. "I just hope the rest of the bloody luggage is here."

I peered over the front seats, trying to get a look.

The tires crunched on white pebbles as we pulled into a driveway.

The house was the ugliest one I'd ever seen. Its walls were olive-green wooden slats, and it was perched on top of concrete posts, which were joined together with faded green wooden boards. The roof was dull grey. The house presided over a withered front garden, where shrivelled weather-beaten plants hung limply. Many, it seemed, had just given up.

"Oh my God!" my mother cried. "What a horrible colour! I thought you said it was cream, not green!"

"That's the place next door, number 13," my father said. "This one was cheaper and it's number 15. Thirteen is bad luck."

"So, you let your stupid bloody superstitions guide you to buy us a God-awful army-green house with a dead garden?"

It wasn't a good start.

Scowling, my father climbed a small set of wooden steps and opened the front door.

"Stop your whingeing and get inside. It's bloody hot out here." Coming back down, he started grabbing bags from the back of the van. Sweat was pooling in his deep frown line.

Mark and I ran up the stairs. We entered a small, open area that led to a long, thin hallway with two doors off it and another door at the end. The cream lino floor was grimy and etched with cracks and splits. The walls were an assortment of creamy hues smudged with scuff marks. The place smelled stale and mouldy.

The house contained some basic furniture: four single beds in the three bedrooms, wardrobes, a red Formica kitchen table and some pink vinyl chairs. In the middle of the lounge room sat a beige car bench-seat, with a couple of small tables and a tiny black-and-white TV.

"This'll have to do until we can get a settee," my father said.

Pointing to the hall, my mother screamed. "Cockroaches!" Several black, shiny roaches scuttled across the lino. I jumped back. I'd never seen ones so big.

"You've always got to find something to whine about, don't you?" my father said.

I saw my mother's eyes squint, her lips purse. A flush crept up her neck to her face.

I grabbed my brother's hand and we ran. Behind us, I heard a thump and a cry. We darted through the kitchen and into a room with a concrete floor. It was cooler here. Another small room off to the side contained a toilet. A pile of battered moving boxes were stacked in a corner.

"This must be the laundry," my brother said. "And our stuff's here."

"At least Dad will be happy," I said.

We sat on the concrete floor until my mother walked in. Her eyes were puffy and there were dried tears on her cheeks.

"Here you are. I was worried," she said. "Sorry for going off like that earlier. I just don't like this house."

"It's okay, Mum," I said, trying to work up a smile.

"We've put your bags in your rooms. Mel, you'll have the one next to ours and Mark has the one near the front door."

We went to unpack. My room had a curtainless window, dull grey carpet and the same musty smell. The single bed had wood-grain paper peeling off its headboard, but its pink foam mattress looked new. The room also contained a plain wooden wardrobe and a battered pale pink dressing table with a small oval mirror.

This was our new Australian home.

It didn't seem quite so bad when I discovered the rear garden. It was huge, dotted with large trees. Some had fruit— nectarines, apricots, peaches and plums. The garden also hugged one side of the house. An enormous gum tree commanded the space. Its colossal, gnarly branches jutted out of a trunk that was too wide to hug.

This would be my new hiding place, I decided, wedged in a nook between the lower branches, reading books.

MONTHS PASSED QUICKLY. I had turned eight years old. I started at the local primary school, my brother at the nearby high school. Kids teased me incessantly about my accent. I desperately wanted to lose it, so I would fit in.

While I missed my sisters and Angie at the centre, I made friends with a girl across the road called Shirley, who went to the same school. Sometimes, she came over after school and shared my nook. She was short and chubby with an upturned nose. She wore her long blonde hair in plaits. She reminded me of Miss Piggy from *The Muppet Show*. I liked her.

My father started his new job in town. My mother seemed to spend much of her time scrubbing the house and complaining about cockroaches. But she also planted vegetables in the backyard and flowers in the front. She grew a choko vine, which produced big green fruits like pears but tasted bitter. She used them in stews. Even in Australia, I couldn't escape my mother's stews. One day, much to my delight, she brought home a ginger kitten. I called him Tiger. I played with him in my room at night, while my parents bickered next door. When Tiger brought in dead mice, my father said he was earning his keep.

But then, things got ugly again.

Mum bought a car and got a part-time job in a nursing home. Now, I'd hear my father demanding to know where she'd been, or why she was home late. She yelled at him, accusing him of being suspicious, and demanding that he stop spying on her and following her home.

One weekend, while I was playing with Tiger out the back, I heard a crash and screams coming from the house. I ran into the kitchen. To my horror, it was like the dreadful scene in Saltash, after my sister Dot came home late, was being replayed all over again.

My mother and father were locked in a grotesque embrace on the floor. Once again, they were wrestling in the middle of chopped vegetables. Our big metal stew pot was lying on its side on the floor. My mother and father were hissing and spitting, their hands around each other's throats. They snarled like angry animals, their words unintelligible.

My mother's eyes rolled to me.

"Help!" she cried.

I ran to Mark's bedroom. He was on his bed, fiddling with his portable radio.

"Quick, I think they're going to kill each other," I said. "Please help."

He looked up. "Leave them to it, Mel. They'll calm down."

"No!" I pleaded. "They're hurting each other. Mark, please come!"

He sighed and got off the bed. I ran ahead of him.

Now, they were upright. My father was pinning my mother to the wall by her arms. A rolling pin was her in hand. She was wailing, calling him useless. She was a slut, he shot back.

I grabbed my father around the waist and pulled. "Stop it, you two!" I screamed.

Mark raced up beside me and, together, we yanked. My father stumbled back, releasing my mother.

He regained his footing and turned to us. We were all panting. His face was flushed with rage, his auburn hair sticking out wildly. His eyes glowed. I was terrified. I heard Mark's voice beside me.

"Look, Dad. Just, just please leave her alone," he stammered. "I know she can be moody, but you shouldn't hurt her."

The vein in my father's neck was pulsating. His mouth was twisted. His hands reached for his waist. He undid his buckle, snapped his belt out of the trouser loops and brought it down on my brother.

"How dare you!" Dad bellowed. "Don't you ever touch me, or tell me how to run my house again!"

Mark raised his arm defensively. Another snap of leather on skin. Mark ran out of the kitchen. My father strode after him. Mark's door slammed; I heard muffled screams.

A short time later, my father emerged, rethreading his belt. He looked dishevelled. He eyed me warily, grunting as he passed.

My mother was picking up bits of vegetables from the floor. She didn't look at me. It was like nothing had happened.

I went to Mark's room. The door was closed. I heard sobbing. I knocked.

"Leave me alone!"

"It's me. Can I come in?"

No response.

I opened the door. He was on the floor, propped against his bed. I noticed red marks on his arms.

"I'm sorry," I said. "I didn't want you to get hurt." I sat next to him. "I was just worried. Next time, I'll call the police."

"Mel, you just have to leave them to it," he said. "You know what they're like. I'm never going to try and stop them again. You do what you want. Things will never change."

In my heart, I knew he was right.

WE FELL BACK INTO the familiar pattern of awful fights followed by uneasy truces. Over the years, I had become good at knowing when the next row between my parents would erupt. Snarls, snaps, sarcastic comments and glares through slitted eyes gave me all the clues I needed. Even Tiger seemed to sense the escalating pressure. He always made himself scarce just before a big blow-up.

Once again, Mark retreated and spent more time in his room. It was like Saltash all over again. Mark said little at the dinner table and avoided my father. I couldn't shake the feeling of guilt. I felt responsible. This time, it was my fault.

When Mark wasn't at school, he was fiddling with bits of old radios on his bed. He saved his pocket money and bought them at thrift stores. He had just turned fourteen and was shooting up like a runner bean.

He confided that he didn't have many friends at school and felt out of place. Girls giggled at him, and boys mocked him for his accent and shyness. The only time I saw him laugh was when we kicked a soccer ball around the back-yard with our new puppy, an adorable brindle terrier called Dougal.

Tiger continued to bring home rats and mice, dumping them at the laundry door. Tiger and Dougal steered clear of each other.

Mum bought me a blue budgie. I named him Blitz. Every morning, he jumped from his perch to his cage door for a scratch on the head. He learned to say "Hello gorgeous" and "Who's a good boy?" I thought he was super-smart.

I wrote to my sisters. They wrote back, telling me stories about who they'd seen, and where they'd been. I always felt so homesick when I read their letters. As time went by, they didn't come as often. One day, the letters didn't come at all.

One thing that did continue was my parents' fighting, with violent eruptions every couple of weeks. Mark closed the door and stayed in his room. I never asked him to help me again, but I couldn't ignore the fights. When it got phys-ical, I would try and break it up. I wedged myself between Mum and Dad, pleading with them to stop. Occasionally, I got whacked in the crossfire or pushed aside by my father, who yelled at me to butt out. But he never whipped me as he had Mark.

My father kept grilling my mother about her movements. She'd yell and poke at his chest with her finger, calling him a "stupid git." They argued a lot about bills and money. When I felt the fights were careering out of control, I called the police. Officers then turned up, separated them and urged them to calm down. The police became my peacekeepers and the only people I could rely on to help. I often spied Shirley and her parents in their yard across the road, looking at the police cars outside our house. I felt so embarrassed.

When my father was in one of his less-volatile moods, he sometimes wanted me to go with him to a block of land he'd bought up the road. According to my mother, he was trying to build a house. He'd pack the car with all sorts of tools, bits of metal and wood, and drive us to the block. While I really wanted to explore the large, empty paddocks and the light-house on the cliff, my father was more intent on using my labouring skills. He asked me to hold this, do that or fetch a certain tool from the car. I did as I was told. I never got to the lighthouse, but at least the block was a change of scenery from home. Sometimes, he even acknowledged my useful-ness. But he also never failed to remind me that I was still hopeless and would never amount to anything important.

School was no escape. I was shy and found it hard to socialise and make friends. I wasn't good at sports—I was slow and uncoordinated. I often feigned illness, so I could avoid going. My mother complained about missing shifts at the nursing home to look after me.

One morning, sitting at the kitchen table, I declared, once again, that I wasn't going to school.

My mother erupted. "You *are* going!" She slapped two pieces of buttered toast in front of me.

"I don't want to," I whined.

Her cheeks flamed. Grabbing my ponytail, she yanked my head back. She picked up a slice of toast and shoved it at my mouth.

"You are bloody well going to school. Open your bloody mouth and eat your breakfast."

I opened my mouth. She shoved the toast in. A bit lodged in my throat. I started coughing violently. I couldn't catch my breath. I wanted to throw up.

"Stop!" I blurted out. Pieces of toast flew across the kitchen. I doubled over, spitting bits and crumbs on the floor.

I looked up at her. In that moment, I despised her. Her eyes widened. She flopped on to a chair next to me, grabbing my hand.

"I'm sorry, Mel," she said. "I just get frustrated. I didn't mean to hurt you."

Tears welled up.

"It's just that your father is so possessive. He follows me everywhere, you know. He wants to know who I've spoken to, who I've seen, where I've been. I don't like this house and I really don't like Adelaide."

I said nothing and looked at the floor.

"I've told you before, my mother died when I was very young. I had to live with my father and look after him. My sister and brother moved out. They were older than I was. If my mother was still alive, I would have left your father years ago."

I felt overwhelmed with conflicting emotions. I felt sorry for her, yet I resented her. I craved her affection and approval but, at the same time, I didn't care. I wanted her love but I didn't want the turmoil that came with it. I knew she cared about Mark and me, but she was also selfish.

She stood and sighed, studying my face. I wiped the remaining crumbs from my mouth.

"You like horses, don't you?"

"Yes." *Black Beauty* was another favourite book.

"Okay. I know you can do riding lessons in the hills. If you like, this weekend, we can go up there. We can enrol you. You can learn how to ride. I'll pay for it. Your father doesn't need to know."

A surge of happiness flowed through me.

"Thanks, Mum," I croaked. My throat still felt sore. "I'll go to school."

MY MOTHER KEPT her promise. That weekend, we drove to a place called Woodside, about 40 kilometres outside of Adelaide. As we travelled through the undulating hills, I realised these were the same mounds I'd seen from the plane approaching Adelaide. The transformation was amazing. They were no longer the dry, bronzed fields; the landscape was green, lush and abundant. It reminded me of my trips to the countryside with my sisters. I was excited.

We arrived at the stables. It wasn't hot, but I felt the subtle sting of sun on my face. The air smelled grassy. I heard whinnying and snorting.

We walked to a dusty yard, where horses of all shapes and sizes were tethered to a wooden railing. They were all saddled and bridled. Their tails swished back and forth. Children were milling around.

A woman greeted us, then fitted me with a blue velvet hard hat. She showed me to my horse. It was mottled grey with a deeply dappled rump. Its big brown eyes and long eyelashes reminded me of the pony I'd met at Dartmoor, but its coat wasn't as thick. I ran my hand slowly along its neck. Its shoulder twitched under my touch.

"This is Jilly," the woman smiled. "She'll be your ride today. If you take more lessons, we'll make sure you get her. It's important that you know your horse and she knows you."

The woman showed how to put my foot in the stirrup, hoist myself up and hold the reins. I watched intently. After several clumsy attempts, I was in the saddle. Jilly didn't flinch.

"Good job," the woman said.

My mother, standing next to her, smiled.

"Jilly's one of those horses that tolerates beginners. She's not flighty. A bit slow, but she won't bolt on you."

I leaned forward and stroked Jilly's mane. I breathed in deeply. She smelled like freshly cut hay. She whinnied softly. My heart soared. I fell immediately and deeply in love.

6

NO CHRISTMAS CHEER

JOCK'S PHYSICAL STRENGTH *always surprises me. Although he is gentle, on occasions his speed, agility and sheer power in the water take my breath away.*

It's a blistering day in the middle of summer. We've headed out extra early, just two hours after dawn.

An amber glow flings ombre beams across the water; heat seems to radiate from everywhere. The hush hanging over everything feels motionless, oppressed. Even the birds are static, perched on mangrove branches like porcelain figurines on a mantelpiece. The water appears lit from within as if by candlelight. The stillness has allowed the sediment to settle, sunlight to penetrate.

We putter into Jock's patch. As usual he greets us with a symphony of raspberries. It is such a stifling day; I am keen to jump into the water, even though it feels as warm as the air above it.

As soon as I slip in, Jock swims away, turns and races towards me.

"What are you doing?" I yell out to him as his fin speeds in my direction. I feel momentary panic. What is he going to do? I trust him but I'm also beginning to understand his mischievous streak, his cheeky spirit and occasional desire to play energetically.

I can do nothing but tread water and trust in him. As he reaches me, he slows a little and thrusts his snout into my chest. He continues swimming at speed, pushing me through the water at a breakneck pace. It's a strange feeling, being pushed backwards through the water so quickly that a pressure wave forms behind me. But I allow myself to be propelled, I surrender control, knowing his agility in this environment is far greater than mine and I know he won't intentionally harm me. On this occasion and other times he chooses to play this game, he never pushes me too far away from the boat but it always reminds me that I am a visitor in his world and he's in charge.

The act of shoving a fully grown human through the water seems to tire Jock out. After releasing me he waits until I get my bearings and swims alongside me back to the boat.

Our playtime ends in the backwaters. Covered by a mangrove canopy, with mottled sunlight flickering on ripples, I relish the feeling of just hanging out with Jock. I flip onto my back and float with the current with Jock beside me. We drift, weightless, his blowhole just touching the skin of the water. He closes his eyes. His regular "puhs" are hypnotic, almost like a metronome keeping time when time itself seems to have stood still. The experience is otherworldly, serene. It's as if Jock and I are just content in one another's company in that moment. We are separate, different species, but it doesn't feel that way. It feels as if we are simply two sentient beings embracing the stillness and peace. I feel untroubled, interconnected with everything around me, allowing myself to be enveloped by the wonder of it all.

It takes a while for me to come back to reality and help Mike find other dolphins. And once again it is a day that shows me the strength of these mammals.

BECAUSE IT IS SO CALM, we decide to head out to the mouth of the Port River, near the breakwater.

The river flows into Gulf St. Vincent, the large inlet of water lined with sugar-white stretches of beach sheltered by gently sloping dunes and ancient sandstone cliffs. The river mouth seems to be the unofficial border separating the Port River dolphins from the other bottlenose dolphins cruising the coast. Sometimes, however, there are unexpected incursions.

An hour or so into our trip, before heading back to check on Jock, we come across a small group of female dolphins. We know them; they frequent the wider channels. Drifting with the current, we watch them for a while. To me, they embody happiness, living peacefully with one another and their environment. They seem content just to hang out with one another, diving and chasing the occasional fish.

From the border, three dark fins slide quietly towards the group.

"I don't recognise those dorsals," I say.

Mike grabs his camera, zooms, clicks and studies the image.

"No, I reckon they could be bikers. Let's see what they do."

"Biker" dolphins, as he calls them, live along the coast. They are generally bigger and a deeper graphite grey than those in the river. Many have scrapes and blemishes, battle scars from propellors, sharks or other rough encounters, sometimes with one another. On the few occasions I've seen them before, they have stayed away from our boat, choosing to linger in small pods just outside the river's entrance.

Now, the trio of bikers hurtle towards the pod of females. They move among and between the females, apparently intent on separating them. The females quickly disperse. Shrill whistles splinter the silence.

"They're bullying them!" I cry.

"I think they're hitting on them. The females don't seem interested though," Mike replies.

To my surprise, the females regroup and approach the bikers. Whack! Whack! Flukes slap the surface. The force of the impact is so strong, it forms foamy white caps. The water bubbles and surges, stirred by a sudden tempest as the females smack and thump their powerful tails.

I've seen this behaviour several times before, when dolphins fish and slap their tails to herd their prey into shallow water, making them easier to catch. Now, they appear to use this action as a unified warning.

I smile. These girls are gutsy. They're laying down boundaries, not tolerating the intruders' behaviour. Eventually, the bikers turn and head back out to sea. The females resume frolicking like nothing has happened.

SOMETIMES, YOU HAVE TO FIGHT BACK. *Nature isn't always kind. I understood that domination is part of the natural order, allowing animals to exert their genetic or perceived superiority. But the natural world also equips its inhabitants with the means to retaliate.*

The females were displaying their strength. They had clearly chosen not to acquiesce to the strangers in their territory. They had stood their ground and protected each other. They had made it clear, without hurting the bikers, that the intruders' behaviour was unacceptable.

I admired their courage. I hadn't always been so brave. I wished I had been more lion-hearted and hadn't cowered to people who believed they were stronger or superior.

I told myself that, from now on, I would take a leaf out of the female dolphins' book. Somehow, I would find the courage to stand up for myself and fight for what I held precious.

IN 1981 I STARTED high school. My friend Shirley from across the road went to the same school. From the moment I stepped through the imposing wrought iron gates of Brighton High, I was bullied. Although I had managed to lose my strong Cornish accent, I still sounded different from the other kids. I pronounced words differently. I'd talked my mother into bleaching my hair blonde to get rid of the auburn tinge and look more Australian. But the kids teased me mercilessly about my bleached hair; even though I lied and told them it was natural, they didn't believe me. Shirley also told them about the frequent visits by police to my house. She laughed and poked fun at me with them. I felt betrayed.

By Term 3, I was spending all my lunchtimes in the art room. No one wanted to hang out with "Dye my hair" or "Horrible Horrill," the girl who always had the cops at her house. I liked art, so I'd sit and draw or make papier-mâché creations. Occasionally, teachers would pop in but, mostly, they left me alone.

When the teachers looked at me, I could see pity in their eyes. I didn't want their sympathy. So I withdrew into myself. I spent more time in the library, reading as much as I could about how to look after horses. My only company at lunchtimes was the librarian, but I managed to avoid her sympathetic gaze.

Most lunchtimes I lost myself in daydreams about one day owning my own horse, galloping across faraway fields. Although life didn't hold much joy, I found happiness in my imagination.

MOST OF THE KIDS at school couldn't wait for Christmas. While I did like it, I also loathed it.

The day always started off well. Presents under the tree, colourful crepe paper decorations strung across the rooms. My mother always made an effort to buy Mark and me a nice gift. A bowl of mixed, salted nuts sitting on the coffee table, the smell of chicken roasting in the oven. The sound of my father grunting and swearing, as he pushed pieces of pork, chicken liver, sage and stale bread through his ancient hand-mincer to make stuffing.

Lunch was always special—stuffed chicken, roast veggies and leg ham, followed by Christmas pudding or trifle. My father had a few light beers, my mother a few sherries. Mark and I were allowed a tipple. We ate at the red Formica kitchen table, which was shrouded with a plastic cloth covered in a bright mistletoe print. We used it every year. We pulled Christmas crackers, and took turns reading the bad jokes. My father refused to wear a paper crown.

After lunch, however, things always changed.

Neither of my parents were heavy drinkers. Their cocktail of choice was twelve months' worth of leftover rage, hatred and frustration, served up steaming on Christmas Day.

After helping Mum clean up, Mark and I usually escaped to our rooms. We knew the afternoon would hit a nasty crescendo. I had a tape deck, so I would sit on my bed losing myself in the Little River Band and, later, Australian Crawl. "Downhearted" became my Christmas theme song.

The afternoon fights were particularly awful; they usually erupted in my parents' bedroom. I'd turn up the volume to maximum to try and drown out the screaming. Inevitably, following the raised voices would come a familiar crash and crack. Another broken lamp, glass or ornament. Sometimes, I'd investigate, knowing full well what I'd see. My mother always seemed to come off worst in Christmas arguments. Blood would be trickling from her nose or mouth, while my father's hands were clenched into fists. When I tried to break up the fights, he'd remind me of my uselessness and stupidity. One year, he threw a pack of cigarettes at me, telling me to kill myself with them. His words burned.

Our only regular visitors at Christmas were police officers, coming to broker an uneasy truce. My father would drive off to work at his block of land, while my mother tended to her wounds.

Nothing more was said. Life continued.

When I was about twelve, however, the Christmas routine changed. After lunch, my father went outside. He was pouring a concrete driveway, and said he wanted to "get on with it" because the weather wasn't too hot. Mark and I looked at each other. He was chewing his bottom lip. I started biting my nails.

That year, my parents had invested in a fish-and-chip shop, in which I worked some nights after school and weekends. Mum had left her job at the nursing home to run it. I helped her prep salads and serve customers. I even learned to flip burgers. My father worked there at night, and Mark helped out occasionally. The shop went broke within six months after a man had found maggots in his barbecued chicken. I knew that, with the failure of the business, my parents had lost a lot of money.

Mark and I went to our rooms. I pulled out my trusty tape deck and put on Australian Crawl, my Christmas song. "Downhearted" blared from the speakers. Tiger lay on the bed next to me, snoozing.

An hour or so passed, the tape had long finished. The only sound was Tiger's purring. I wondered if we'd finally have a peaceful Christmas.

I decided to go outside; maybe I'd sit in my tree nook and read for a while. Book in hand, I walked through the house to the back door.

I heard a scream. It wasn't my mother.

As I stepped outside, I saw my mother and father on the lawn. He was shrieking and wailing like a wounded animal. My mother was kneeling on his chest. One of her hands was attempting to pin down his arms above his head, her other hand was between his legs.

I ran towards them. She looked up. She grinned a hateful grimace.

"I got him," she said. She was squeezing between his legs. He was thrashing, his eyes closed, face contorted, tears flowing down his cheeks.

"For God's sake, you crazy bitch!" he shouted. "Let go."

"Here, Mel. Help me hold him down." My mother sounded breathless. She stared at me, her eyes blazing; they looked wild. Her lips were twisted, her nostrils flared.

"What?" I said. "Are you bloody bonkers?"

My father's head thrashed from side to side, spit was dribbling from his mouth.

"Just do it," she demanded.

I dropped to my knees and threw my weight on top of my father's pinned arms. She kept squeezing between his legs.

"Got you for once where it hurts, you useless bastard. What kind of man loses money like that? No balls, that's your problem."

He wriggled and writhed. I felt the strength of his arms under my weight.

Finally, he sat upright. I flew across the grass, landing face-down. On her knees, my mother eyed him warily as he scrambled to his feet. He stood, holding his crotch, looking down at her. I'd never seen his face so crimson. It looked ready to combust. I curled into a ball, bracing myself for the explosion. I looked from one to the other. It was like a scene from a movie. I wasn't really there—I was observing this from another place.

His eyes radiated a savage glow as he looked down at my mother. Her mad grin had vanished; her face was now vacant, emotionless.

"You've gone too far." He spat on the grass. "You and your child need to get the hell out of my sight."

She stood and scuffled over to me, yanking me to my feet. She dragged me inside. I felt lightheaded, woozy. She took me to my room and I collapsed on the bed. Tiger was still on it. Mum dropped next to me. She said nothing, but reached for my tape player. We sat with Tiger and listened to the Bee Gees. Nothing was said, but the air felt laden with dread.

That night, she dragged her single foam mattress into my room. She slept on the floor next to my bed, for which I was grateful. I lay awake, feeling soiled, dirty and somehow permanently tainted. I couldn't get the image of my mother's sneer out of my mind. I knew I shouldn't have helped her, but I had. I was ashamed of myself. I realized I had crossed a line. I was no longer just a witness to and vocal protestor against the violence; I had actively participated in it. I had committed it. I hated myself for it.

Eventually, I dozed off, only to wake abruptly in the dark, my heart pounding, overcome with the familiar desire to run. All I could hear was my mother's deep breathing. I concentrated on the slow rhythm of her breaths and, somehow, I fell back to sleep.

My mother slept in my room for quite some time. Most nights, she dragged in the foam mattress and slept next to me. Even though I was now a little afraid of her, I was more frightened of my father—and a little afraid of myself. I knew mum cared about me and wouldn't hurt me on purpose. I also sensed that she needed my company as much as I needed hers.

I BEGAN WAKING at night more often. I always woke with a start, terrified, my heart beating rapidly, with an overpowering need to flee. My mother called it "night terrors." I wondered if evil demons were hunting me, like Jadis, the White Witch of Narnia. I tried to lose myself in books, but it was hard to concentrate and I couldn't conjure the mystical worlds as vividly as I used to. I felt numb, empty. I knew something deep inside me had changed, but I wasn't sure what. That scared me even more.

I escaped from home whenever I could. I rode my bike to the paddocks near our house to sit and watch the horses. I desperately wanted one of my own.

Every second weekend, I looked forward to riding lessons. To help pay for them, I started a Sunday job, spinning cotton candy at an amusement park. Sometimes, I got shifts on Saturdays and long weekends. I gave my paycheques to my mother. I eventually got my brother a job there, attending the giant waterslide.

At the riding school, I was upgraded to a more spirited horse called Thor. He was a stocky, dark brown gelding, an ex-pacer. While he could be stubborn at times, he trotted beautifully.

One day, when I was putting my gear away, I noticed a large black-and-white sign on a noticeboard.

"Give away to good home. Eleven-year-old bay gelding in poor condition. Needs retraining. Good temperament." On the bottom was written a phone number with the name Gail.

I snatched the paper from the board and ran out to my mother, who was waiting by the car.

"Mum!" I cried, waving the paper in the air. "Someone's giving away a horse!"

I desperately wanted to convince her. She studied the notice.

"It won't cost anything; I can pay for his feed with my job. He can stay at that place up the road. It's only ten minutes away. I'll look after him. I'll ride my bike there every morning before school and feed him. I can retrain him. I'm a good rider now. It's my birthday soon."

She glanced up from the paper.

"I don't know, Mel. It's a big responsibility. We'd have to buy the equipment."

I was bursting with excitement. "They sell second-hand tack here! We can get a saddle, bridle and bits. They even have old rugs. I'll pay you back. I'll do extra shifts over the school holidays. Please!"

"Tell you what," she said. "I'll use the phone here and give them a call. If this place is in the hills, and this Gail woman is around, we can look. But no promises. It could be an old nag with health problems. Okay?"

I threw my arms around Mum's waist. "Thanks, Mum!"

My mother walked into a small wooden building, which housed the office. I waited outside. I couldn't stand still. I nibbled my nails, kicked the dirt, twirled my ponytail around my finger. I felt butterflies in my stomach. What was taking so long?

"Alright," she said, as she emerged. "It's not far from here. I've got the address."

I couldn't get to our old red Leyland P76 quickly enough. We wound through hilly backroads until we found the property. The small paddock had a dilapidated shed in the corner. We pulled up in front of it. There was a handwritten cardboard sign on the barbed wire fence: "Horse to give away to good home."

Standing in the middle of the sparse, dusty paddock was a large horse. Head down, he was nibbling at thin tufts of grass. He was a deep, rich cinnamon colour with a sable mane and tail. I could make out ribs protruding from his side. A bony hump bulged on its rump.

There was a crunch of wheels on gravel as a car pulled up. A woman who appeared to be in her early twenties got out. Her wispy blonde hair was tied in a high ponytail. She had a thin face and pale lips, which she pulled into a smile when she greeted us.

"Hi, you must be Doreen." She held out her hand to my mother. "I'm Gail."

"Yes, hello. And this is Melody, my daughter."

I smiled.

"Let's go and see Eddie," she said. She walked up to a rusted metal gate. We followed her in.

Eddie looked up, eyeing us warily.

"Hello, Ed," Gail called. "Come on."

Eddie sauntered slowly towards us. From the books I'd read, I saw he didn't have good conformation. In other words, he wasn't put together well. He had a large head, out of proportion to his body. His shoulder blades poked out. His unshod hooves were cracked and jagged. I'd learned you can judge a horse's character by their eyes. If you could see white, the horse was skittish or mean.

As Eddie stopped next to Gail, I could see his deep cocoa eyes looked gentle, with no white. They were curtained by long, black lashes. I wished I had some carrots in my pocket.

"I haven't ridden him for ages, maybe a year," Gail said, running her hand along Eddie's neck absently. "I'm just too busy. My boyfriend doesn't like horses. I just want him to go to a good home."

I felt a pang of sympathy. "He looks so thin."

"He needs worming. Plus, there's not much fodder left in this paddock. He's a good boy, about eleven, I reckon. He's just over fifteen hands high, so not too tall, but tall enough. He rides nicely and isn't skittish. He'd make a good first horse with a bit of retraining."

I held my upturned hand to Eddie's nose. He sniffed and snorted. His upper lip started twitching and twisting, then a huge pink tongue popped out and licked my palm. I was stunned. It was warm, damp and rough. It was the strangest feeling, like a sand-filled burlap bag that'd been lying in the sun being dragged across my fingers. He kept licking.

"He seems to like you," Gail said. "Either that, or you've been handling something salty."

I looked up at my mother, agape, watching Eddie's thick tongue lap at my palm.

"Okay," Mum said. "Can you keep him until we pick him up? We'll have to hire a trailer."

My heart leaped.

"Sure," said Gail. "You have my number. Give me a call and I'll help you load him." She looked at me. "Well, you've got a new friend. Promise me you'll take care of him and be kind. He may not be the most attractive boy, but he's got a gentle heart."

I nodded. I knew Eddie and I would be best friends.

WHEN MUM TOLD DAD, he was furious.

"A bloody horse! What's she going to do, charge for kids to ride it at the beach?"

I rarely, if ever, tried to speak with him alone, but I approached him one evening, when he was fiddling with his van. I felt huge rain moths flapping around in my tummy, bumping into my stomach wall.

"Dad, I know you don't want me to get a horse," I said quietly. "I don't want to make money off him. He won't cost you anything, I promise. I'll work as many hours as I can to cover his costs."

He wiped his greasy hands on his pants and studied my face.

"Another bloody mouth to feed, and it's not even going to earn its keep. It'll eat money, that's what it'll do."

"Dad, please. I've never asked you for anything."

He stared at me. I couldn't read his eyes. I held my breath.

"Whatever," he said. "Your mother won't get off my back until her spoilt child gets what she wants, so to hell with it. I don't want to know. But the minute he gets too expensive, he's gone."

He turned back to his van.

7

JOY AND HEARTBREAK

AS MONTHS ROLL BY, *I grow to understand more about Jock and myself. I watch Mike jump in the water with him and notice that he seems more at ease with this wild dolphin in the river than at any other time. I get the feeling that he cares deeply for Jock and at the same time is learning a lot about dolphin behaviour and intelligence.*

I too am discovering more about my capacity to trust and love. I am not a scientist but I know that Jock is conscious and aware, and I hope he is getting as much out of our friendship as I am.

Jock follows the boat into the deeper waters of one of the larger channels. I'm wearing a wetsuit as winter has decided to intrude for a day or two. I don't like wearing a wetsuit, although it helps keep me afloat. I find it harder to move and I feel somehow disconnected from the river. Feeling the water on my skin is something I have grown to love. But I also know that I can spend longer in deeper, cooler water if I wear a suit. I decide against the flippers.

On this morning, I descend into Jock's realm and manoeuvre myself upright, treading water, bobbing erratically, like some unstable ocean buoy. I wonder what Jock is in the mood for today—playing hide and seek, playing with the paddle or maybe a bit of tag? Jock blows a raspberry and sidles up to me. To my surprise, he doesn't do anything but lie still, his blowhole just above the waterline.

I speak to him as I caress the side of his silky body, revelling in the cool, smooth sensation and marvelling at the contours of the muscles connected to his tail fluke, which propel him through the water. I run my palm over his scarred mouth, once again feeling the indentations left by embedded fishing hooks and feel a stab of sadness at his suffering. I feel his rough snout, wondering what he's eaten that morning and hoping it was something better than crustaceans in sediment. He takes a breath and rolls over, and his eyes meet mine. Yes, there is intelligence and awareness in them. I carry on stroking him and speaking to him softly. He seems to be just mooching, grooving in the moment and enjoying the attention.

He rolls onto his back. His tummy is exposed. It's pale, almost eggshell pink and delicate. I can see blue veins under the surface—pumping blood to his heart and organs, like in my own body. It strikes me that although we are different species, maybe we aren't that different after all.

The tips of his pectoral fins pierce the surface. My hands move across his belly and I gently tickle the area where his torso and his fins meet. I am completely oblivious to the sounds of the river—crying gulls, distant boat motors, the hum of the power station. It feels surreal. How can this be happening? This wild creature is letting me tickle him and he seems to be enjoying it.

Minutes pass. Jock flips onto his front again, takes a breath and swims slowly back to the boat.

"He let me tickle him under his pectorals," I cry as I clamber back on board the research boat. Mike nods and smiles in response. I have never felt so close to Jock before. I'm starting to feel my armour fall away. Was I finally allowing myself to feel vulnerable? If I could feel this way about a dolphin, could I begin to open up my damaged heart to others?

My emotional horizons are widening as is my view of the world around me. It's hard to describe the exhilaration at feeling so close to the natural world and its rhythms.

A few months later I am to experience another thrilling day, this time on a yacht heading across Gulf St. Vincent beyond the mouth of the Port River.

I'M FLYING, ON MY STOMACH. Leaning over the bow of a yacht, my hand is dangling over the edge, centimetres above the water; it's damp with sea-spray. I think if I can reach down a little farther, I can touch the swift bichrome dolphins leaping and spinning beneath me. But they're just out of reach and we're travelling too fast.

Overhead, the great white spinnaker is bulging and brimming as it propels us forward. The yacht splices the surface, like a razor cutting through paper. The sea is butterflying beneath me—the sharp bow sculpts rolls of turquoise and white, curling away to oblivion. There's no motor running, just the thunderous sound of water being sliced and diced, and the blustery winds clanging through the rigging. I feel intoxicated by the sea, speed, whoosh of the water and the throng of dolphins surrounding me. It's almost too much. Too magical. Too exhilarating.

I'm on board Independence, a magnificent yacht owned by Mike's friend Paul, crossing the gulf. Our destination is Kangaroo Island. We have hours to travel, thank goodness. I've taken many trips across the gulf before, usually just to Port Vincent, an eight-hour trip using wind power. Today, we're going farther on a four-day return journey.

I wriggle myself away from my bird's eye view, turn over clumsily and sit up. Paul is standing at the stern, his hands lightly touching an oversized metal wheel fitted with a compass. The steering wheel seems way too big for the job. I wonder, do yacht-makers do that on purpose so the person sailing can feel like a master, in command?

Mike is snapping away as usual, immersed in his task. Paul calls in my direction. "Hey Mel, wanna give it a crack?"

I've sailed this yacht under Paul's tutelage a few times before. I love feeling the pressure of the water resisting the carve of the rudder. The sensation travels up through the hull and into the wheel, which feels alive. The wheel, however, responds to gentle touch. Paul says you can sense the yacht as well as sail her.

I get up and move to the stern and slide in front of the wheel as Paul steps away. The wind is behind us. Good. I look up at the pregnant spinnaker. I'll hold my course. I know we'll have to tack soon, which will require all of us to help. Sailing is a complicated, physical business.

After a while, Paul takes the helm back and we change direction. He barks orders. We release the spinnaker, point into the wind, and hoist the mainsail. Duck under the boom and alter course.

The slight change in our trajectory is like an invisible beacon, calling more dolphins to join us.

On the horizon, dozens of dual-toned silver missiles zoom towards us. Hurtling at startling speed, spending more time in the air than in the water, they seem to be skimming across the surface, like the stones Mark and I used to skim as kids, zig-zagging alongside, under and in front of the yacht. The sea boils with life.

These are common dolphins. The top half of their compact bodies is charcoal-coloured, the bottom ash, like an engineer has painted them to be more aerodynamic. Their home is farther offshore than the bottlenose dolphins back in the Port River. Unlike the Port River dolphins or the bikers cruising the coast, they all seem perfect, flawless and lustrous, as swift as supersonic jets.

Like their river counterparts, these dolphins seem to want to join us for the fun and thrill of it. As always, their joy is contagious. It isn't just the dolphins' unconditional companionship and infectious zest that prompts our elation. It's their sense of freedom, the serenity of being concerned with nothing but the sea. Of being connected to something authentic, real and vast.

Eventually, we make our way into American River on Kangaroo Island and drop anchor. That balmy evening, we lie on the deck, looking up at the stars. The universe appears so black, yet so illuminated and interminable.

I doze, rocking languidly in the arms of my equally fathomless mentor—the sea—lulled by the occasional "puh." While my solitary dolphin friend Jock is always on my mind, here, for a while, it is so peaceful. I wonder if, by learning from nature and my growing love of animals, I might possibly find a way to experience some real peace in my own life.

THE WEEKEND AFTER meeting Eddie, Mum, Mark and I hooked up a rented horse trailer to the tow bar of the P76, and set off to collect him. I still had a while to wait to turn sixteen and be able to drive myself.

We'd already visited a property near us, and the owner had agreed to agist Eddie for a small monthly fee. It was the same paddock where I'd often ridden my bike to watch horses.

Gail was there to greet us. She led Eddie into the trailer. He didn't hesitate, but clambered up the ramp and stood still as she tethered his halter to a rail at the front. She said her farewells. I was amazed she didn't cry.

An hour later, we pulled up at Eddie's new home. I went into the trailer, untied him and gently guided him backwards down the ramp. I took his halter and led him into a small, fenced-off yard, where he would stay on his own while we got him checked by a vet, reshod and wormed. The occupants in the main paddock stopped grazing and stared at us. A couple of them whinnied.

"It'll be okay, Ed," I whispered in his ear, which twitched at the sound of my voice. "You'll be out there with them soon."

The vet came while I was at school. Mum said she had to hold Eddie as the man pushed a huge tube of worm paste into his mouth. I knew Mum was a bit scared of horses, but she held on to him anyway. The vet said Eddie was in good health but malnourished, and suggested we feed him oats mixed with molasses and bits of apple, along with his usual lucerne hay. Mum made a deal with our greengrocer to buy old or damaged apples cheaply. We purchased litres of thick, black molasses in brown barrels and big burlap bags of oats.

The vet's diet worked wonders. Within a few weeks, Eddie had gained weight. The hump on his rump became less noticeable and I could no longer see his ribs. His eyes were brighter. His hooves looked smooth and shiny, and were now fitted with shoes.

I had little interest in school. I'd count the minutes until I could leave, get on my bike and spend time with Eddie.

His training came along nicely. Since putting on weight, he had more vigour. I rode around the paddock in circles, urging him to trot, canter and come to a stop when I pulled gently on the reins. I talked to him the whole time. His ears, now covered with a velvety fuzz and tufts of fur, flicked back and forth. He was listening to me.

On cold and wet winter mornings, my mother who, surprisingly, seemed to be enjoying this new adventure, would get up extra early and drive me to the stables. These were basically large, open sheds with rickety roofs. I carried a bucket filled with oats, bran and molasses, mixed with steaming water, to warm him up. He trotted up to us from across the paddock, vapour streaming from his huge nostrils. He always licked my hand. Mum and I would readjust his raincoat as he scoffed down his breakfast. I saw Mum smiling; she was starting to like Eddie and feel less afraid of him. This made me happy.

On lighter evenings and school holidays, I took Eddie for long rides. Not far from his paddock was a vast parcel of vacant council land. All the owners rode there, along the winding fire-trails that stretched for kilometres.

As we galloped up hills, wind whistling past my ears, I revelled in feeling his strong legs propel us forward and upward. Wisps of mane whipped my face as I leaned forward, urging him on, telling him how wonderful he was. Warm air scented with cut grass rushed against my cheek. The dull thudding of hooves hitting the ground filled my ears. The feeling of total freedom was exhilarating. It felt like Eddie and I were one. We flew together and nothing else mattered.

Back at the stables, Eddie would be panting and wet with sweat. I'd hose us both down. He'd close his eyes as I poured cool water over his back, then over my head. I'd always give him a treat: a sugar cube, carrot or piece of turnip. I'd stay until it was almost dark, brushing him, telling him how much I adored him.

On hot days, I rode him bareback down to a nearby beach called Seacliff, with a small empty bucket attached to my backpack. It took us about half an hour to walk there. I rode him along the sand and into the sea, never going deeper than my thighs, being conscious of sharks. Water lapped at my legs and his sides. I closed my eyes, taking in the briny tang of the air, the distant sound of waves lapping at the shore and children laughing.

Sometimes, Eddie licked the water with his rosy tongue as he glided across the sandy bottom. Beachgoers stopped and pointed at us. A girl in shorts and bikini top, riding bareback through the sea, obviously wasn't a common sight.

Out of the water I would dismount and on a long rein, he'd drop to the sand and start rolling in it, his long, lanky legs in the air, hooves skyward, his body writhing and wiggling backwards and forwards. He looked in ecstasy. Bathers stared and smiled. Children came up to me, asking if they could pat him.

After lolling in the sand, he'd stand and shake himself, like a giant leggy dog who'd been revelling in something smelly. I then led him up the concrete ramp and filled my small bucket with fresh water. I sploshed water over him until most of the sand had washed off and gave him a drink.

At that moment, life couldn't get any better. I never wanted to go home. I wanted to curl up in his stable and sleep next to him, breathing in his earthy, horsey smell.

BACK TO SCHOOL. Brighton High was renowned for its music program. I took up the bass guitar but lost interest after a few months. I joined the choir instead. Shirley barely acknowledged me anymore. She hung out with the pretty, loud, popular girls. I continued to spend lunchtimes alone in the art room or library. It was close to the end of my second year at the school.

I'd endured constant teasing about my hair and family from many of the girls and a few of the boys in my year. Others just avoided me. If they were aware of my home life, the teachers never let on. I wasn't sure if they knew, but I reckoned they had an inkling. There was always that hint of sadness and sympathy in their eyes when they looked at me. They never asked me about it and I never brought it up.

One afternoon, I was waiting for the singing teacher in the choir room, along with more than a dozen other students. The room was cavernous with a high, steepled ceiling. A shiny black piano sat in one corner next to a conductor's stand. Light streamed in from the glass doors, giving the room an angelic quality, which seemed appropriate. Rows of orange plastic chairs with metal legs faced the front. Some were filled with students.

I wiggled on my seat, closest to the door in the first row. I always chose the front row because it seemed safer somehow. There were empty seats next to me. I felt the familiar tingle of stares piercing the back of my head. Girls were giggling and whispering. I recognised voices, the same ones that taunted me mercilessly about my hair and police visits to my house.

I felt a hard shove at my shoulders. I fell off the chair, face forward, crashing to the floor. Hands started pulling my hair, pinching my arms, yanking at my cotton dress. There seemed to be hundreds of them. I rolled into a ball, my arms covering my head. I squeezed my eyes shut. Muffled screeches, chuckling and babbling enveloped me. I heard myself scream, pleading with whoever was attacking me to stop. I felt something sharp slide down my leg.

"Enough!" a male voice boomed. "Get back to your chairs, now!"

The sound of metal legs scraping against the floor.

Dazed, I unravelled myself. I rose, felt for my chair and pulled myself back onto it. I looked at my lap. A pocket dangled limply off the side of my dress. I examined my leg. A long scratch extended down the length of my calf. It wasn't bleeding. Then I noticed bits of hair lying limply on the floor. I choked down tears.

My music teacher loomed over me. I looked up. His black eyes bored into mine.

"I think you need to go home, Melody," he said gently. "I'll get the office to call your mother."

I stood, wobbly, and walked out of the room. Silence trailed behind me.

I didn't look back.

A short time later, my mother picked me up. As we walked from the office, she eyed me over. We got into her car. She didn't start the engine.

"What happened?" She turned to me, slumped in the passenger seat. "The school said you'd got into trouble and suggested I come and pick you up."

"A group of girls, they always pick on me. Some of them jumped me in the choir room. I should've fought back."

"And what good would that have done you? How many were there?"

"I dunno, maybe six or seven."

"Well, be thankful you got off as lightly as you did. They could've broken something. Where was the teacher in all this?"

"It happened before he got there," I replied. "Please, can we just go?"

With a deep sigh, my mother drove out through the school gates.

"I don't want to go back," I mumbled. "Please, don't send me back."

"Where the bleeding heck are you going to go then?" she said. "Your brother's school is bloody terrible. Where the hell am I supposed to send you?"

"I don't know, Mum. I'm sorry. I just want to go somewhere the kids are nice. I don't care, I just can't go back there."

She glanced at me, and a low groan rumbled in her throat. "Okay, let's see. I'll look around. I can't promise anything, and I certainly can't tell your father about this."

A week or so later, my mother told me I'd be going to a private college in the city. She had persuaded the principal to waive the fees. It was a secretarial secondary school; I'd learn how to type and do shorthand along with normal subjects. I'd catch the bus there and back.

Two months later, I started at Muirden.

My father, who was spending more time at the block of land at Marino, didn't notice my new routine or different uniform. I felt relieved. I didn't want another confrontation.

The school was small, wedged on the corner of a busy city road and a side street. It didn't have vast, beautifully manicured grounds like those at Brighton. It didn't have a sports field. It was a dull, grey, three-storey building. It could easily have been mistaken for an office block.

Immediately I noticed the kids there seemed different. They appeared a bit rougher and less refined, louder and more boisterous. They didn't comment on my bleached hair.

Some of the other students had brightly coloured streaks in their own hair. One of the boys even had bright carrot-orange tips that stood out at random angles. He also had a small silver ring in his ear and another through his nose.

Many of my fellow students, I discovered, had come from broken families, which relied on the principal's generosity to help with school fees. For the first time, I felt I might belong somewhere.

The principal was a tall skeleton of a man with a bald head, yellowing teeth and rounded shoulders. He looked old. He wore the same black suit and tie every day and spoke slowly. He smiled kindly at me when he saw me in the corridors.

I met Jan during my first few weeks there. She was new, too. She was much shorter than I was and slightly chubby. Her fine, wavy copper hair ended at her shoulders. Huge sapphire-blue eyes dominated her pale oval face. They were the most beautiful eyes I'd ever seen. I couldn't stop staring at them. Her two front teeth were crooked. We became friends instantly.

I confided in Jan, describing my adventures with Eddie and traumas at home. I sometimes showed her the bruises I collected during interventions. She told me about the fighting between her own mum and dad. She said it sometimes got violent, but was nowhere near as bad as with my parents.

We smoked cigarettes in the girls' washroom together, shared hot chips smothered in gravy and rode the same bus. She lived farther south than I did. My stop was on the way.

I STARTED ENJOYING some of my subjects, especially typing and shorthand. My favourite teacher, Mrs. Erskine, taught English.

"Melody, can you wait a moment, please?" she asked while I was packing up my books after class one day. Jan shot me a worried look as she left the room.

I stood at my wooden desk as she walked over. She struck me as being young for a teacher, slightly taller than I was and slim. She wore half-moon glasses, perched at the end of her pinched nose. Her short, lank blonde hair was always tucked severely behind her ears. She seemed to have an endless cupboard full of tartan skirts and wore a different one every day.

"Look, Melody, you're doing very well at English," she said, peering over her glasses. "You were the only student I gave an A-plus to on their report card last term."

"Really?" I remembered the swell of pride I'd felt when I first saw the score. A handwritten comment also said I was a pleasure to teach, and had a bright future if I applied myself. I recall Mum being pleased.

"I like English. You're a good teacher." I smiled. "You're about the only teacher I've actually ever liked."

"Yes, well." Her cheeks coloured. "Thank you, that's kind. Look, I'm going to nominate that piece you wrote on the poems of Wilfred Owen for the annual school English award."

I was speechless. I'd read Owen's poems over and over. I felt deep sorrow as he described being on the battleground during the First World War. I imagined his anguish at the futility of war and the horrors he had witnessed in 1914. I had used his book of poetry as the subject of an essay on early-twentieth-century writers.

"I can sense things aren't very good at home," she said gently. "But I think you've got a real talent. Maybe you should consider a career in journalism. I also think you should buy some chewing gum; your breath reeks of cigarettes."

It was my turn to blush.

A month later, Mrs. Erskine presented me with a small silver spoon in front of the other students at assembly. Its handle was embellished with the school's crest. My name and the year were etched in cursive writing on its scoop. I'd never won anything before and was overcome with gratitude.

I tried to dampen my father's words, which still echoed in my head. Maybe I wasn't stupid and useless. For the first time, maybe I could start believing in myself. Maybe.

I CONTINUED AT MUIRDEN, working at the amusement park on weekends, and sometimes after school and during the holidays. I earned decent money, which I handed to my mother every week to help pay for Eddie. I spent every spare moment at the paddocks with him. Mark was also still working at the park, enjoying his job supervising the waterslide.

Then came sadness. Tiger died, run over by a car. I was inconsolable. A few months later, we lost our terrier, Dougal, after my father unbelievably left him in the car on a searingly hot day. We buried them both in the backyard. I'd lost two dear friends and companions.

I had a relationship with a boy that lasted for a few months, but we broke up when he told me he still liked his ex-girlfriend. I was heartbroken, I thought, but the feeling didn't last long.

My brother, Mark, had bought himself a car and sometimes drove me to see Eddie. A few times, he let me drive on private land. One day, I backed his car over a well on the property and it hung precariously on the edge until we got a tow-truck. He was reluctant to let me drive again.

Things at home didn't change. Police were still regular visitors. Sometimes, I saw Shirley watching our house from across the road. I didn't speak to her anymore.

My father complained increasingly about Eddie. I heard him telling my mother that the money I earned should go towards my keep and the household, not on the horse.

ONE AFTERNOON, about eighteen months after adopting Eddie, I arrived home from school to find my mother sitting at the kitchen table with a cup of tea. Her eyes looked puffy.

"Sit down for a minute, Mel," she said.

I dropped my school bag and sat next to her. She took my hand.

"Look, we're going to have to sell Eddie. I'm sorry. I know you love him, but it has to be done."

"What?" I shouted. "This is Dad, isn't it? He's got to you, hasn't he?"

I snatched my hand away. "I pay for everything," I cried. "Okay, you sometimes help with the farrier, but I give you everything I earn!"

"That money has to go into the house, Mel. You know I'm not doing as many shifts at the nursing home now and things are tight."

"No! I'll quit school and get a full-time job, or I'll do more shifts at the park. I give you more than enough for his food and agistment. No, Mum, please. We can't sell Eddie."

I felt the sting of tears. She looked at the table, and started picking and poking at a crack in the corner. I noticed her hair looked good. Maybe she'd had it cut. I felt a surge of anger.

"Look, your father and I agree on this. I've already put an ad in Saturday's paper. If we get any calls, I'll go with you on the weekend. We'll make sure he goes to a good home. Maybe you can keep the saddle and bridle? You might get another horse one day."

My simmering anger erupted. I leaped from my chair. "So, we can't afford Eddie, but you can afford to go to the bloody hairdressers?"

I grabbed the back of the chair and tossed it aside. It crashed to the floor.

"I hate you!" Tears rushed down my face. "I can't wait to leave." I glared at her. "And never sleep in my room again. You and that bastard deserve each other!"

I ran to my room, slammed the door and threw myself onto my bed. I buried my face in the pillow and howled. I reached for my little worn Jody and hugged him tightly.

A knock at the door. "Mel, are you okay?"

"Go away, Mark! They're going to sell Eddie. Just leave me alone."

He did.

The rest of the week was a blur. I spent as much time as I could with Eddie, staying at the paddock until dark. I avoided my mother, father and Mark.

SATURDAY MORNING CAME.

The air felt thick and damp, unusually humid for Adelaide. Fat, dark clouds squatted in the sky, about ready to burst.

The phone rang. I heard my mother pick it up. I crept up the hall so I could listen.

"Yes, you can come this morning," I heard her say. "Yes, he's in good condition and would make a great first horse... He's my daughter's... Melody's almost sixteen... No, we're keeping the saddle, but you can take the rug... Yes, he's shod... No, $400 is what we're asking."

She explained where the paddock was. As she replaced the receiver on the wall-mounted phone, I snuck up behind her. She turned and jumped.

"God! Don't scare me like that," she said putting her hand on her chest. "You obviously heard what I was saying. I'll drive us there. They want you to ride Eddie for them. I told them we'd be there at 11:30." Her tone was matter-of-fact, cold.

I didn't respond. I walked to my room, closed the door and sat on my bed. I felt my emotions brewing inside me. I tried to tell myself Mum would change her mind when we got there and she'd say I could keep Eddie.

An hour or so later, my mother came into my room. She was still in her dressing gown, rollers and slippers.

"Mel, I've got one of my migraines coming on. You'll have to go there by yourself. I'd ask Mark but he's at work."

I stared at her blankly. Migraines were a fairly regular complaint of hers, along with painful varicose veins, high blood pressure and swollen feet.

"Don't think of shirking out," she continued. "There'll be hell to pay if we don't sell him. And don't lie and say he's a bad horse. I'll find out."

She left the room.

No, I wouldn't lie, but I wasn't about to do a sales pitch.

During the bike ride there, I told myself the people wouldn't like Eddie. He'd be too big, or too ugly. He wasn't a pretty horse; he was out of proportion. That would put them off. They were only coming to look, that's all. Once they'd seen him, they'd change their mind.

The air was clammy and oppressive. The heavy sombre sky felt sinister, like the whole universe was going to collapse and suffocate me.

When I arrived, I greeted Eddie and gave him a handful of hay and chopped apples. He licked my hand. I saddled him up, led him to the dirt road at the bottom of the paddock and waited. A car, towing a red horse trailer, pulled in. It stopped when it reached me. A man and woman got out. They looked to be in their late thirties. They were dressed in jeans and T-shirts.

"Hi, you must be Melody," the man said. He didn't smile. He had short, dirty blonde hair and ratty eyes. "This must be Eddie." He nodded towards him.

"Yep." I responded flatly.

The woman standing next to him said nothing.

"Can you jump on and show us how he rides?"

I did as he asked. I urged Eddie into a canter up the hill to his paddock, then trotted him back down. I repeated this several times, feeling terrified that this would be our last ride together.

When I felt Eddie was getting tired, we walked back to the waiting couple. They were eyeballing him, looking at his legs and his face. They whispered to each other and nodded.

The woman went to the car and retrieved something from inside.

The man said, "So, your mother assures us he's a good horse. Is he skittish?"

"No, he's never bolted on me and he doesn't shy easily." I caressed Eddie's neck.

The woman sidled up to the man. He took something from her.

"Okay," he said. "Here you go." He handed me a piece of paper. "We'll load him up as soon as you unsaddle."

My jaw dropped. "What? You're taking him now?"

"Yep, I think he'll do nicely."

My mouth felt dry. I looked at the piece of paper. It was a cheque for $400.

I undid the saddle, then gently removed Eddie's bridle, replacing it with a rope halter the man had handed me.

"Thanks," he said, and grabbed the halter. He started leading Eddie towards the trailer.

I watched, rooted to the spot. "Wait!" I cried.

The man stopped. I ran to Eddie and threw my arms around his neck, nuzzling my face in his mane.

"I will never forget you, Ed," I whispered. "One day, I'll find you and buy you back."

I released him. The man turned and loaded him into the trailer. I watched in disbelief as they drove away. I examined the cheque in my hand. I'd just exchanged my horse for this piece of paper. I shoved it in my jeans pocket.

Ed's saddle was lying on the ground. In a daze, I picked it up and carried it to the stable. Then I got on my bike and rode home. The heavens opened. Fat drops of cold rain. It started pouring. I was getting saturated, but I didn't care.

When I arrived home, I threw my bike onto the back lawn and stormed inside, dripping, heading straight for my parents' bedroom.

I charged in, hair plastered to my head, drops of water staining the carpet. My mother was lying on her bed with a towel over her eyes.

"There you go." I threw the cheque on her chest. "He's gone. I hope you're happy now. Here's your precious bloody money."

She removed the towel and opened her eyes, glancing down at the cheque.

I ran outside and sat in my tree nook, sobbing. I didn't bother getting changed. I didn't care if I caught pneumonia.

Thunder rumbled in the distance. I knew it wasn't safe to be in the tree during a storm, but I wanted a bolt of lightning to come and fry my pain.

For weeks, I moped around the house, asking my mother repeatedly to get Eddie back. She told me she'd already cashed the cheque, and couldn't track down the new owners anyway. I didn't believe her. Nothing I did came close to filling the void in my heart. I felt so lonely, so hollow.

I thought the anguish would never end.

These close-ups show how this beautiful, powerful mammal and I connected closely. Miraculously, Jock let me and other research assistants explore his dorsal fin and back. His terrible scars around his mouth were the result of discarded fishing hooks which had become embedded in his gums. His rough snout was most likely the result of rifling in the silty riverbed for food. MIKE BOSSLEY

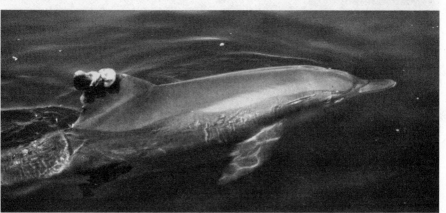

top: Jock, like other dolphins I'd seen, seemed to enjoy leaping in the waves created by boats. While most dolphins jump and play in bow waves at the front of boats, Jock appeared to prefer springing up in the frothy wake behind the research boat. MARTIN JACKA

bottom: Jock's distinctive dorsal fin was mangled from discarded fishing lines and nets. His disfigurement allowed us to recognise him instantly. MELODY HORRILL

One of our first games with Jock involved a boat paddle. The top picture shows fellow research assistant Steve and me playing with this favourite toy of Jock's. He always seemed to know what we were trying to do, and carried off his trophy! I was constantly amazed by Jock's love of playing games. Many times, I had trouble keeping up with him, but he seemed to realise this and slowed down so I could catch up. MIKE BOSSLEY

top: I loved watching Jock leap high behind the research boat. Eventually, we used this game to help lure him away from his solitary home and meet other dolphins. MIKE BOSSLEY

bottom: Being propelled backwards by a powerful dolphin is an unnerving experience. But, somehow, I knew Jock wouldn't hurt me. His strength was amazing and, when he wanted to play this game, all I could do was let it happen! MIKE BOSSLEY

8

THE PENALTY
FOR PLAYING

OCCASIONALLY JOCK FOLLOWS US to the invisible border of his territory. Mike opens up the throttle as the channels widen and deepen. Jock seems to love it. I gasp at his ability to keep up with us.

Behind the boat's frothy wake Jock appears in mid-air. He flings himself completely out of the water, landing with a belly flop back into the gurgling, fizzy bubbles before leaping out again. Over and over, he launches. I wonder if he feels a sting as I do, landing in the water on my belly. If he does, it doesn't seem to bother him. I am always concerned that he'll lose the top of his dorsal if he smacks down too hard. As Jock leaps higher and higher, I also worry that he'll misjudge and land in the boat. He never does.

Martin Jacka, a local newspaper photographer who occasionally comes out with us, becomes very excited at Jock's airborne antics. Martin stands in the back of the boat, throws his hands in the air and urges Jock to leap ever-more skyward.

"Woo hoo, Jock!" Martin screams as he grabs his camera and snaps the airborne dolphin. In his photos, Jock seems to be flying, his scarred smile even appears somehow more genuine.

Mike and I laugh at Martin's wild enthusiasm but all three of us are euphoric. Watching Jock leap clear of the water over and over without any apparent effort is astounding. It prompts feelings of pure delight and wonder in me. Jock perhaps represents a sense of freedom we all secretly long for.

As time goes on, Jock follows us slightly farther and farther out. But there is always a point where he stops, turns around and returns to his home.

But we keep going. I realise I am beginning to see the river and its inhabitants with new eyes. Perhaps being with a photographer like Martin and seeing Jock through his eyes allows me to observe my surroundings in a different way.

THE RIVER ITSELF triggers a physiological reaction in me. Every time I'm on the water, especially when Jock is around, a happy switch in my brain turns on, a bit like I had first felt with Eddie. I often feel euphoric. The environment around me intensifies, becoming sharper, brighter. I notice more.

Although on top of the natural food chain, the dolphins live in harmony with all of it. In fact, they seem happy to work with other species.

When not keeping my eyes peeled for dolphins, I discern details about other life around me, which I would have once ignored. This ecosystem is constantly moving and evolving. It isn't stuck in time. It doesn't dwell on what it was yesterday. What's important is the present, and how this would morph and grow into another day, another cycle. I feel humbled by this revelation, and ashamed of my own reluctance to leave the past behind me fully and transform into something new.

However, I am beginning to understand my own metamorphosis. I hope I can emerge as agile and rejuvenated as the life around me.

I observe the different hues of aquatic vegetation: the bright green of sea lettuce, the muted forest tones of seagrass, the subtle differences in the shape of fleshy mangrove leaves.

I am in awe of the cormorants as they duck-dive only to appear seconds later with small, writhing fish. The egrets, their alabaster feathers flawless, are shockingly vivid against the water's subtle palette. How majestic they look as they survey their kingdom from a solitary twig.

When steely skies make it difficult to locate grey dolphins, we often look out for the birds. At times, it seems like every seabird species chooses to descend on a single square metre of water. From a distance, it's chaos. The air fills with squawks, shrieks and cries—a discordant orchestra tuning up before a concert. A bedlam of cormorants, gulls and egrets, all swooping, diving and flapping in a jumble of wings, beaks and lanky legs.

In the middle of the mayhem comes the silent, gentle emergence of fins. A harmonious, graceful, coordinated fishing expedition. Loud inhales are followed by tail flukes vanishing, diving into the deep. Moments later, the sudden flash of burnished bullets as spooked juvenile fish jump skyward, perhaps sprayed by sonar. In a raucous crescendo above the surface comes the frenzied rush of birds, all vying for a snack at once.

But this is not mayhem, it is synchronicity. The dolphins are feeding themselves, helping each other hunt. They don't seem bothered by the birds. They have no influence over the commotion, the anarchy around them. They are simply focusing on what they have to do together to survive.

A FEW MONTHS after I sold Eddie, I turned sixteen. I could no longer escape to the fields with him. I missed him every day. I felt as if my heart would never mend. I needed another way to find freedom, another means of escape. I wanted a car.

Mark had forgiven me for the well incident in the paddock and offered to teach me how to drive. I mastered most things quickly, but had trouble learning how to release the clutch. I always did it too quickly. We bunny-hopped around the back streets of Brighton.

One day, Mark said we should tackle a main road. I was nervous. We stopped at a busy intersection. Then the light turned green. I put the car into gear and tried releasing the clutch. The car lurched forward and stopped. I tried again and again, becoming increasingly frustrated. I started swearing. The car stalled. A racket of horns blared behind us.

"For God's sake, Mel. Just ease the clutch out smoothly. How many times do I have to tell you!" Mark said.

He unwound the window, stuck his arm out and waved to the traffic, which had banked up behind us. He grinned at the passing drivers, who stared back, threw a hand, made rude gestures and mouthed words I was pleased I couldn't hear.

I knew Mark was getting fed up with playing instructor, but he didn't give up on me. We spent Saturday afternoons at empty shopping centres. He taught me how to parallel park. I laughed at the look of shock on his face, after I performed it successfully on my first go.

"I'm good spatially," I laughed, feeling surprised and proud of myself.

He smiled. "I'm impressed. My little sis knows how to fit big things into small spaces. That's quite a skill."

I leaned over and hugged him. A week later, I passed my driving test.

My brother knew a lot about cars. He worked on his own for hours on end. He offered to help me find my own wheels. I'd managed to save $100, and my mother gave me $300. I suspected it was from Eddie's money.

I found a twelve-year-old pumpkin orange Leyland Marina, which I bought from an elderly lady who lived not far from us. It drove like a boat with a broken rudder. I'd turn the steering wheel, but the car took a while to respond. When it finally did, it lunged around corners rather than turned.

Despite its flaws, I liked it. I sometimes drove it to Eddie's paddock. I sat looking at his field, remembering our times together. My heart felt hollow as I wondered where he was. But I hoped he was happy and was eating lots of apples and carrots. As much as I enjoyed the freedom of having a car, I'd have given anything to have Eddie back.

One afternoon, as I was turning into the driveway at home with Mark in the passenger's seat, I misjudged the opening. Panicking, I hit the accelerator instead of the brake. We jolted forwards violently. There was a loud crash. Our journey ended on the front lawn, surrounded by bits of cement sheet. My parents were not impressed.

Mark said he'd try to fix the front fender, which was crinkled like an empty chip bag. He spent hours on his back, under the car. He bashed the crumples with a rubber mallet. He slathered it with foul-smelling goop then painted it with orange spray paint from the hardware store. The fender still looked like a bowl of lumpy porridge, but the damage was less obvious. I hugged Mark fiercely and told him he was the best.

Soon after, I traded the Marina at a local car yard for an old Mini. It was dirty brown with a worn gold racing stripe and faded bronze mag wheels. I'd been eyeing it for a while; it looked sporty. I thought Minis were cute. They were made in England, so they must be okay. My mother gave me $200 to cover the price gap.

Even though we were underage, I escaped from the house and drove Jan and myself to the pub on Saturday nights. Before heading out, we'd spend hours in my room, trying on treasures we'd bought from thrift stores, or poring over the impossibly pretty girls in the latest teen magazine. We carefully applied blue eyeshadow, thick black eyeliner and shiny red lip gloss on each other.

One afternoon, when Jan and I were getting ready to go out, she was sitting on my bed while I slathered her hair with copious amounts of spray, trying to make it look like Pat Benatar's in the music video for "Love Is a Battlefield."

"You've got such pretty hair," I said, fluffing and squirting. "Mine's dead from years of bleaching."

She looked at me. "You're the prettiest girl I know, Mel. And you're so smart. I know your dad doesn't think so, but you are."

I was touched. I kissed her on the cheek.

Jan was shy, careful and refined. She was completely opposite to me in many ways. I was often blunt, coarse and tactless, and I'd swear loudly. She rarely raised her voice.

"All the boys like you," she said. "I'm fat and ugly. They only speak to us because of you, I know that."

I stopped spraying, grabbed her shoulders and turned her towards me.

"No, you're not," I said sternly. "You have the most beautiful eyes, they're like sapphires. You don't have any pimples like me and your smile makes the world okay. I don't know anyone else who can do that. You're the best friend I've ever had. And you're also much better at math than me."

The bedroom door flew open. My father stood with hands on hips, glaring at us.

"You look like a pair of floozies," he said. "Off to tart around again tonight, are we?"

We didn't respond. I turned my attention back to Jan's hair. Eventually, he trudged off.

Later, with my father off at his block of land, my mother asked where we were going. I suspected she knew about our underage clubbing, but she didn't seem to mind.

Our favourite spot was a big hotel on the other side of town with a nightclub on the second floor. There was always a line-up at the door. We both looked older than sixteen so no one questioned our age. We became regulars and one of the bouncers, a stocky young man with thinning blonde hair, started to let us in for free. He winked and smiled at me. I thanked him using the name on his badge. Duncan. He responded with a sheepish grin.

The club was cavernous. The stench of stale tobacco hung in the air. A bar brimming with spirits lined an entire wall. It reminded me of a pharmacy, its shelves stacked with strangely shaped bottles filled with coloured concoctions, which promised to cure every ailment.

Dominating the centre of the room was a sunken wooden dancefloor as big as a football field. Silver-legged, terracotta-coloured chairs and tables hugged its perimeter. Disco balls and lights hung from black rafters on the ceiling, creating dappled rainbow rays that flittered and shimmied across the expanse. It was like entering another realm where everyone was happy. Boys with quiffs and mullets, wearing thin leather ties and bright jackets, strutted alongside girls with one-shouldered dresses and implausibly high hair, defying gravity.

I loved to dance and would always be the first on the dancefloor with Jan. We squealed when Duran Duran, Icehouse and Mondo Rock came on. The floor filled with people swaying and bouncing in circles around piles of handbags, drinks in one hand, cigarettes in the other.

The music swaddled me. I allowed myself to be lost in the throb of the rhythm and the otherworldly surroundings. Occasionally, we'd talk to boys and dance with them. They were always a passing distraction, however, not to be taken seriously. Jan and I chuckled at their clumsy advances.

Jan was always by my side. If I found a boy creepy, she'd agree. If I told someone to get lost, she'd stand with me and politely tell him to leave us alone.

We'd stop at drive-through liquor stores on the way to buy small bottles of vodka that we'd smuggle in. We bought glasses of Coke from the bar, slipped into the ladies' room, tipped out some of the Coke and replaced it with the vodka. We saw lots of other girls doing the same with tiny bottles of rum, brandy and scotch. We often got tipsy, but never drunk.

One evening, we got into the Mini to go home. As I drove out of the hotel's parking lot, I heard a loud clunk from under the car. Without warning, I felt resistance from the gear stick. It crunched and grated. I couldn't hold the wheel and move the stick. There was too much resistance. Something was terribly wrong.

"Jan, it's too hard for me to change gears. You're going to have to do it while I drive."

"I've never driven!" she exclaimed. "You're going to have to tell me what to do."

"Just put both your hands on the gear stick and, when I say 'change,' you move it down, then up to the right and down again, okay?"

She frowned at me. "Okay."

We drove home at a snail's pace, Jan grunting with exertion as she forced the gnashing, grinding gear stick into place on my command. I kept both hands on the wheel, worried about being stopped by the police. We pulled into the driveway forty minutes later.

The following day, I asked my brother to look at the car. He was training to be a nurse but told me he would look at it when he could. Eventually he discovered that the clutch plate needed replacing; he could probably fix it. It took him days to do, but he did it.

As a thank you, I bought him a box of chocolates.

My getaway car was back in business.

ONE NIGHT, WHEN I pulled the Mini into the driveway, I noticed a figure lurking in the dark.

"Who's that?" Jan asked nervously from the seat next to me.

"Not sure," I replied. "I think it's probably my father."

I opened my door cautiously and got out. A dazzling light blinded me. A torrent of something cold and wet battered me. I lost my footing and slipped onto my bottom. I heard Jan screaming. I looked up to see a stream of water ricocheting off her chest.

"What the hell?" I yelled, scrambling to my feet.

The water stopped. The light dimmed. My father's figure strode out of the gloom. He was holding a spotlight in his left hand, a hose in his right.

"You two dirty bitches have been out whoring again, haven't you?"

I heard the venom in his voice. He moved closer. "You're just like your mother."

He turned to Jan, her saturated shirt clinging to her skin.

"And you," he snarled. "You're just as bad. Get out of my sight. You're not welcome in this house."

"Stop it, Neil." My mother's voice boomed out of nowhere. "They're just having fun, no need to give them a chill. It's bloody freezing out here!"

She emerged from the darkness and walked over to Jan. "I'll get you a towel and give you a lift home. Melody, get inside and get changed."

"Sorry, Jan," I said, sullenly.

Jan screamed at my parents. "You're both insane, you know that?" She pointed to my mother. "You're a crazy witch." She turned to my father. "And you're an evil lunatic."

I held my breath. "Don't bother about driving me home," Jan added. "I'll call my mum from the phone booth up the road." She turned and stomped off.

I was livid. I didn't know whether or not to run after Jan. I decided it was probably easier to let her be.

"See what you've done!" I ran inside.

I saw Jan at school the following Monday. She apologised for yelling at my parents. I told her not to worry about it. She said she didn't want to visit me at home anymore. I told her I didn't blame her. We agreed I'd visit her and we would still go out together after things cooled off.

ABOUT A WEEK or so later, I was lounging on my bed one evening, flicking through a copy of a teen magazine which Jan had snuck into my school bag that day. I could hear the rustle of my tree against the window.

Then I heard my mother and father start arguing in their room next door. I grabbed my tape player, popped in an Australian Crawl cassette and turned up the volume.

Through the wall, I heard my father's disjointed words. "Your daughter... no better than you."

"Always jealous... controlling git."

There was an almighty crash and bang against the wall. I felt my bed shudder. Then a high-pitched scream. This sounded bad. I'd never carried a weapon before, but something told me this time I might need one.

I picked up an old metal hand-mirror from my dressing table. Armed, I tiptoed to their room, creaked open the door and snuck my nose around.

My mother was on her bed, back slumped against the wall, head lolling to one side. A broken wooden coat hanger lay beside her. There was a nasty red line on her cheek. A thin stream of scarlet trickled from her nostril. Emotion had vacated her features.

My father was sitting opposite her, on his bed. Jaw in his hands, he seemed mesmerised by the blood sliding down her chin. His shirt was hanging out of his trousers. Tousled greying auburn waves pasted to his forehead.

I hesitated, wondering whether I should call the police again. Something inside me snapped. A surge of intense anger gushed through me. I strode up to him.

"You bastard, you made me sell my horse. He was the only thing I cared about. I'm sick of it, I'm sick of you! I know you hate me but I don't care." The words spewed out of me like vomit. I couldn't contain them.

He looked up and scowled. I raised my metal mirror and whacked it down on him. I heard it shatter as it met his shoulder. Shards of glass flew across the room. I dropped the frame on the floor.

"What the—?" he cried. He leaped from the bed, grabbed the front of my sweatshirt and drew me up to him. He wasn't much taller than I was. I noticed his frown line as he pushed his face to mine. I could feel his breath. His raw clay eyes drilled into me. They seemed to seethe with hatred.

"How dare you!" he bellowed. "You'll be sorry."

I said nothing. I just stared at him.

"You're useless," he continued. "You'll do nothing with your life, you're too stupid. You'll end up pregnant, with a swarm of bastard children, like your mother. You're nothing!"

His words didn't sting. I'd heard them too many times before. I didn't react. This enraged him further.

"So, you've got nothing to say? You're not even my daughter—you know that, don't you? You shouldn't even be here, I'm sterile. None of you are mine!"

None of this made sense. He'd told me numerous times I wasn't his but now he was claiming he couldn't have children. Was this just another sadistic lie?

I felt a sharp sting across my face. I heard the thwack of his palm hitting my cheek. I stumbled. Glass clinked under my feet. I backed out of the room slowly, holding my hand to my burning cheek. I didn't take my eyes off him.

Once clear of the door, I ran through the house and out the back door, snatching my car keys from the kitchen counter on the way. I fumbled to unlock the car. I got in and pushed the key into the ignition, but didn't turn it. Instead, hands shaking on the steering wheel, I stared into space. My cheek was on fire. I realised I couldn't leave without knowing Mum was okay. But I couldn't bring myself to go back in. I was crippled by indecision.

Minutes, maybe hours, went by. I heard a knock at the window. It was Mum. I opened the door and got out. Together, we trudged to the back doorstep and sat on it. Blitz was chatting away to himself.

I studied her face. The rivulet of blood was starting to cake. The red welt was ballooning.

"He hit you. I'm sorry, Mel."

"It's okay, Mum," I said, rubbing her arm. "It was just a slap."

"You broke your mirror," she said, a small smile twitching at the undamaged side of her lip. "He's going to hate that. It's seven years' bad luck. You know how superstitious he is."

"Where is he?" I asked.

"He's still in the room." Her voice was slurred. Her gaze shifted to the ground.

I shuffled onto my knees and settled in front of her.

"Look at me," I said.

She raised her face and scrutinised my eyes.

"You have to leave him," I said. "WE have to leave him. Mark's going to be living at the hospital soon for his nurse training, anyway. We can't stay here, Mum. Someone is going to get killed. We have to go . . . soon."

She nodded slowly. "I know."

At that moment, she looked like a little girl, lost, alone and bewildered. A wave of empathy flowed through me. Despite everything, she was my mother, but I felt like our roles would forever be reversed. I would have to look after her.

I knew what we needed to do.

MY BROTHER MOVED into the residential wing of the Royal Adelaide Hospital to complete the practical part of his nurse training. I knew he didn't really want to do nursing, but my father had been pressuring him for months to find a job and get out. I'd always thought Mark would end up as a mechanic or an engineer. He was so good at fixing things and understanding how they operated. But I also knew he had great empathy and his future patients would benefit from that fine quality.

The house felt empty without him. I envied his freedom.

My mother and I started talking about our own escape.

In a few weeks, I'd be finishing Grade 11. She suggested we leave soon after that, when my father was at work. I agreed. We couldn't go when he was at home or at the block.

Mum said she'd look for a cheap place to live, short term. Meanwhile, we'd start packing bits and pieces, and stash them in bags under my bed. My father never did any housework, so he wouldn't find it.

I made Mum promise she wouldn't provoke him again before we left. She said she'd behave.

We talked the plan through regularly in my bedroom at night before she retired to their room.

One evening, a couple of weeks later, she told me she'd found a small, two-bedroom, partly furnished flat at Glenelg, a few suburbs away. To my surprise, she also confided that she'd made a new friend. She had answered a personal ad in the newspaper and was corresponding with a Sri Lankan accountant called Sid. He was much older than she was and lived about forty minutes away at a place called Seaton. She'd told Sid about our situation, and he'd told her he'd help if she wanted. The news of mum's new friend confused me. It seemed odd that we hadn't even left and mum was already seeking a new companion. I brushed it off, thinking that maybe she just needed some support. I knew she didn't have any close friends—maybe this was her way of finding new ones. I also knew that if my father found out, there was no telling the magnitude of rage that would be unleashed.

Our planned getaway day soon arrived. It was a perfect summer's morning. Bright dappled rays shimmered through my bedroom window. My big tree was silent. Blackbirds chirped.

I dressed quickly.

I walked into the kitchen. Mum was making herself a cup of tea. I sat at the table and started nibbling at a slice of toast she put in front of me. My stomach was in knots. I wasn't hungry.

My father appeared, dressed in his work gear. He took his keys from the counter and walked out of the back door. He didn't acknowledge us. I heard the deep clatter of a diesel engine—he'd traded his van for an orange pickup truck with a flatbed a few months prior.

My mother sipped her tea and I pecked at the toast. The wall clock ticked.

"Right," my mother finally said. "Let's get to it then."

Mum's car was parked in the driveway. We started loading bags into the trunk, along with pots and pans, knives and forks, some plates and a few ornaments that had somehow survived the war.

I put Blitz's cage on an old towel in the back seat. He was chirping and chattering. I left the window open for him.

We were loading the last of our things into the car when I heard a familiar rattle. My father's truck turned into the driveway, pulling up behind Mum's car.

My heart thudded. A feeling of dread enveloped me.

"Mum!" I shouted. She emerged from the door with a stack of sheets. Her eyes widened. She came down the steps and stood next to me.

"Oh no," I heard her mumble.

My father emerged from the car. He seemed bigger and bulkier. His eyes were blazing. He marched towards us.

"What the hell do you think you're doing?" His tone was savage. I knew what was coming.

"We're leaving," she replied, her voice quavering.

"Like hell you are. Over my bloody dead body."

"Just move your car," she ordered shakily. She placed the stack of sheets in the back seat.

My father's mouth twisted. "Make me."

They launched at each other. I raced back to the house, opened the door and lunged for the phone. I dialled the emergency line. My hand was trembling as I tried to poke my finger into the small plastic circle.

"Please," I said into the mouthpiece. "My mum and dad, I think they're going to kill each other. Please hurry!" I gave the operator our address. I ran to the kitchen, grabbed a butter knife from the drawer, and sprinted outside.

My parents were lying on the concrete driveway, wrestling, grunting and swearing, their limbs entangled.

"Enough!" I screamed as loudly as I could. "Dad, stop it. I have a knife!" As soon as I'd said it, I realised the futility of my threat. I couldn't stab anyone, never mind with a butter knife. Who was I kidding? However, my words seemed to have some effect. My mother broke free, grunted and scrambled to her feet. My father did the same. They stood facing each other, staring, eyes wild, sweating and panting.

I heard the wail of a siren. I ran to the street. The police had arrived.

The officers' faces seemed familiar. Two young men. They looked barely out of their teens.

One of them looked at me, and glanced down at the knife in my hand. "You're Melody, the one who made the call?" he asked calmly.

"Yes," I replied.

"What's with the knife?"

I felt suddenly foolish. "Nothing," I lied. "I was making a sandwich."

The officer threw me a sideways glance. The pair walked up the driveway, stopping a short distance from my parents.

"Mr. Horrill, step away from Mrs. Horrill," one of them ordered. My father did as he was told.

"Miss Horrill," he said, turning to me. "Please go inside and put that knife away."

One of the officers guided my mother to the other side of the driveway. The other escorted my father to the rear of the house.

I walked to the front door, placed the knife on the floor and hovered behind the flyscreen. I kept out of sight, but I could hear my mother and the officer talking.

"Did he hit you? Do you want to press charges?"

"Look, we just want to leave," she choked, sobbing. "I don't want any trouble. I just want to go."

"May I ask, where exactly where are you going?"

"A flat in Glenelg."

"What's the address?" The officer took out a small note-pad and pen from his shirt pocket.

My mother reeled off the address. The officer scribbled on the pad.

The other officer appeared, and whispered something to his colleague. They both looked at my mother.

The other officer started talking.

"Look, your husband just wants to know where you're going. You're taking his daughter, after all. I think he has a right to know."

My heart stopped. No, he can't know! I wished I were telepathic, so I could urge my mum to say not to tell him. I wondered if I should go out and demand it.

"Look," the officer continued. "He says he just wants to make sure your daughter is okay. I think he'll calm down and leave you alone, once we give him the address. It's a peaceful way to resolve this situation. He's promised to move his car and let you leave."

My mother sighed. "Alright, but he needs to leave us alone. I don't want to see him again."

"That's a matter for you and your husband. We'll order him to move his vehicle. We'll stay here and watch him until you leave."

"Melody," my mother called. "Let's go."

I came out of the house, looked at the officers and said, "You shouldn't tell him where we're going."

Neither of them responded. They walked off and came back moments later, my father walking between them. His eyes were fixed to the ground.

He didn't look at us, but started his pickup and backed it out of the driveway onto the road.

My mother and I got into her car and reversed out. I turned around as we drove down the street. I saw my father sitting at in the driver's seat of his truck, staring after us. The officers were standing on the footpath, watching.

Although we could run, I feared that we couldn't hide.

9

STALKING

NIGHTMARE

I OFTEN WONDER why Jock trusts us, considering the injuries humans have inflicted on him both intentionally and through carelessness. I can't understand why he doesn't see people as a source of pain rather than pleasure.

Jock is eliciting such joy in me and teaching me so much. He seems to accept me unconditionally, despite my species being responsible for his pain, disfigurement and (possibly) his lone-liness and inability to socialise. It seems so unfair and unjust.

In stark contrast to his treatment by people, to my amaze-ment he displays remarkable trust and concern for my welfare.

One morning, after greeting us with his usual "puh" and raspberry we head into one of the smaller channels between the mangroves. Jock's knobbly dorsal follows. He hangs around the rear of the boat, banging the ladder, eventually unhinging it with his persistent badgering. I grab my goggles, throw my leg over the stern and descend the rungs. The water is warm, as usual, and I greet Jock with a tickle under his chin before slowly dog-paddling away from the boat.

My goggles start to take on water, so I move closer to the riverbank where I can touch the silty bottom, clean my lenses and pop them back on.

As I approach, I pull myself upright. I can feel the squishy mud envelop my feet. I fiddle with my annoying eyewear as I stand waist high in the water.

Suddenly, with a "whoosh," Jock darts towards me and dives. I wonder if he's suddenly spied a tasty mackerel near the bank and has gone to chase it.

Then I feel something very peculiar under me. A rough object is rummaging under my soles, incessantly trying to lift my feet out of the ooze. In an instant, I know it is Jock's snout.

He wedges it under one arch, then the other. Each time I lift my foot and put it back down, he resumes his ferreting. The only telltale sign he's down there is the sonic buzz from his sonar reverberating through my legs and the occasional air bubble rising and popping on the surface.

Is this a new game? I chuckle to myself. Some dolphin version of Twister where, like the party game, I'm not allowed to have two feet on the same spot?

"Mike," I yell. "I think Jock's trying to push my feet up. Is he just playing?"

"No, Melo, he probably thinks you're drowning. He's trying to save you. In the sea dead things sink."

On hearing this, I immediately lift my feet and start treading water.

Jock emerges with a "puh" and quietly sidles up beside me. I am overcome with gratitude. I have never experienced what seems to be such unconditional care and concern. This feeling is foreign to me. I feel like crying; I feel humbled and elated all at once.

Jock hangs in the water, blowhole just breaking the surface. I circle one of my arms around him and rest my head near *his dorsal fin while with my other hand I keep myself afloat. I can hear the clicks of his sonar slowing. They sound like a heartbeat.*

"Thank you, Jock," I whisper. I so desperately want him to understand.

Again, I find myself wondering about this dolphin. How had he become orphaned? I've watched devoted dolphin mothers teach and nurture their offspring. I've witnessed some of them grow into healthy young adults. While these experiences often make me reflect on my own parents, I can't believe that Jock's would have abandoned him.

I wonder if perhaps Jock's mother had been struck by a boat or entangled in discarded line or nets and drowned. Maybe it was pollution or some other form of human carelessness. Perhaps, like Jock, she had been speared but, unlike him, hadn't survived.

I don't want to get out of the water and back on the boat. I want to stay and protect Jock. I wish I could grow a tail and live here with my best friend. I like to think he wants me to stay too. I know it's impossible, though, and I have to return to my own world eventually.

Mike's voice breaks through my reverie, insisting that we need to go. Grudgingly I climb back on board and we head out to find other dolphins.

Within a couple of hours we've seen at least a dozen dolphins. Mike is documenting who we come across and where and recording notes on their behaviour while I ensure we avoid the shallows and sandbars.

Our mood is buoyant as we turn and head back towards the ramp.

Turning into the channel, we notice a solitary dorsal fin, with what looks like a plastic boat fender floating in front of it. Cautiously, we approach.

As we get closer, we realise it's a small body. A baby dolphin, perhaps a few months old. It's on its side, bobbing on the wavelets, one lifeless eye staring at the sky. It looks like a discarded plastic pool toy. Its mother is pushing it. She is nudging it with her nose, lifting her baby's lifeless form above the surface in what appears to be a vain attempt to help it breathe.

The body is adrift at the mercy of the current. The mother keeps prodding, bumping, poking, lifting. It seems like she's trying to will it back to life. The fixed smile on her face makes a mockery of the agony I imagine she is feeling. Her breathing sounds irregular and erratic, coming as a low mournful whistle. I can hear her distress. I can feel it tug at my heart. I can't continue watching, so I go into the cabin, sit down and cry.

Mike's face appears. "You okay?"

"No. Not really. God, it's awful. What are we supposed to do? Are you going to collect the body?"

He clambers down into the driver's seat. "We can't at the moment. We should leave her with her grief. Maybe, once the body gets washed up, we can come back and get it."

"Is she trying to bring it back to life?"

"Dolphin mothers are amazing. If a baby dies, they sometimes carry or push the bodies around for days, even weeks. It looks like this mother has broken off from her pod to do this."

I wonder how the baby has died. Has it been hit by a boat? Poisoned by pollution? Some chemicals can travel through a mother's milk. Have humans caused this suffering? Or has the baby been ill? I also know that a percentage of dolphin babies never survive weaning. Whatever the reason, to me, it seems so cruel, so unjust.

Slowly, we back away, leaving the mother to her fruitless quest. She would eventually return to her pod. She might have another calf. This is the cycle. I wonder again what had happened to Jock's mother.

The experience makes me think; nothing in this world lasts forever. Nature demands that everything will end. Nothing is spared. Then, there is rebirth. There is rejuvenation. We are all a part of this organic process. If we disconnect from this understanding, we disconnect from ourselves.

Watching this devoted mother, with her desire and determination to stay with her dead baby, I think once again of my own mother. I realise I've spent years denying my hidden resentments towards her. While I love her and cherish the memories of loving moments with her, there's also a part of me that is still angry with her for not protecting us children well enough, for not putting our needs as children or mine as a young woman ahead of her own.

I feel selfish thinking this way. I want to believe my mother has tried her best. But, at certain times in my life, when I was afraid and insecure, she focused on her own future, and didn't seem concerned about mine.

Maybe she felt I was strong, that I had it together, but nothing was further from the truth. While I can't change the past, I can make a promise to myself. I will try hard to support and protect those I love and those who love me, whether animal or human. I won't always succeed, but abandoning those I care about isn't an option. I would not perpetuate the pattern.

Watching the dolphins makes me acutely aware of my propensity to dwell on past pain. I sometimes feel trapped by my own story, unable to shake my father's voice from my head. I struggle with negative self-talk. I know I need to let it go.

The dolphins bring light and optimism. They seem to revel in freedom and appear to have innate joie de vivre. Having fun and playing are a big part of their daily lives. They also teach me harsh lessons about the nature of life, death and grief.

OUR NEW FLAT was unremarkable in every way. Cream brick, in a row of six, it sat back slightly from a busy main road. The front door faced the street, and was fitted with a rusty, torn flyscreen. A large window dominated the facade.

A solitary, scraggly wattle tree in front of the window offered minimal shade to the lounge, which contained a worn ruddy-coloured velvet couch with a matching recliner. The décor was sombre and cheerless. The faded walls, perhaps once a rich buttermilk, reminded me of melted vanilla ice cream. Two small bedrooms of a similar size contained single beds and drab chipboard wardrobes. The tiny bathroom had a bulky turquoise sink and white toilet. The shower recess was surrounded by a plastic curtain printed with washed-out sailing boats going nowhere. The kitchenette contained a compact wall oven, its interior smeared with the caked, sooty remains of previous meals. The rancid smell reminded me of the deep fat-fryer from the fish-and-chip shop, after it hadn't been cleaned for a week. A single sink and a few bone-coloured cupboards completed the fit-out. The flat didn't have a garden; instead, there was a strip of asphalt out the front for resident parking.

My mother bought a second-hand TV and a teak table to put it on. This was our new home, but it didn't feel secure. I couldn't shake the fear that my father may know where we were. Despite being surrounded by solid brick walls, I felt exposed.

The move also meant I had to leave Muirden College. Although I had really wanted to start Grade 12, continue with English and maybe pursue a career in journalism, we couldn't afford it. We needed money to live and I needed to help make it. My mother couldn't support us both on her wage, so returning to Muirden wasn't an option. I dropped out; journalism became a distant pipe-dream. Even though I doubted I was smart enough to do it, I had wanted to try. The thought of not returning to school was depressing.

A few months earlier, I had quit the amusement park, after they slowly started cutting my hours when I turned sixteen. I found a job waitressing at a restaurant in the city during lunchtimes. I got extra hours for cleaning its windows after the midday rush, as well as some evening shifts. The tips were much better at night. Once again, I lied about my age.

My mother and I didn't spend much time together. She was either working shifts as a carer or going to see her new friend, Sid. When she was at home, I was working.

When we did see each other, we never discussed what had happened at Brighton. Sid, this strange new entrant in our world, dominated the conversation. She described his beautiful roses, which he manicured immaculately; his lovely new townhouse; and his plans to travel to Noosa, an exotic, tropical, well-heeled town in northern Australia. I listened, but kept telling her I was worried about Dad knowing where we lived. She said she was concerned, but just hoped he'd stay away.

A MONTH OR SO LATER, the nightmare began.

I was sitting on the couch, reading a book, when I felt the hairs rise on the back of my neck. I had the strange sensation of being watched. I got up and checked every room. Nothing. I was alone with just Blitz for company. Creeping to the front window, I pulled aside the net curtain.

An orange flatbed truck was parked on the road outside. The driver's window was wound down. At the wheel was a man in sunglasses and a baseball cap. I thought I could make out a deep frown line on his forehead.

I froze. My heart started hammering in my chest. I couldn't look away. It was like seeing an apparition. I felt he could see my eyes through the small gap in the curtain. I held my breath. I don't know how many minutes passed.

He sped off.

I slumped on the couch. My mind was racing. Should I call the police? He hadn't done anything, so what would I say? It looked like his truck, but maybe there were other orange pickups out there. Maybe it wasn't really him? I had no evidence. If it was him, maybe this was just a one-off. Maybe he just wanted to see where we were living and wouldn't be back. I concentrated on inhaling long, deep breaths and felt my heartbeat slow. It'll be okay, I told myself. You don't scare me; you won't scare me.

But the episode did scare me.

I was so rattled, I called in sick to work. I decided to tell Mum when she got home. I waited anxiously on the couch until she arrived. I heard a car door slam. I looked through the window. It was her. I unlocked the door and ushered her inside before locking the door again.

"I think Dad was here today. I saw him. I'm sure it was him."

"What are you talking about?" She eyed me warily.

"It was his car. He had a cap on, but it was him."

Her brow was furrowed. "Did he get out? Did he come to the door?"

"No, he just parked on the road. He didn't stay long. I think he saw me."

She plonked down on the arm of the couch, and ran her hands through her hair.

"Right. Well, we both need to check the locks before we go out or go to bed. I was going to see Sid tonight, but I'll stay with you instead."

"Do you think he'll be back?"

"You never know with your father," she said. "I hope not, but better keep an eye out. I don't think he'll be stupid enough to try and break in. But I'll let the police know, anyway."

She pulled me to her. "It'll be okay, Mel. I love you."

I stood up and hugged her, twirling her curls around my finger. This was the kind, caring mother I loved and craved for. But her mood could change in a heartbeat. She was a chameleon. For years, I'd slowly built a rickety wall around my heart to humans which was slowly morphing into an impenetrable fortress. No one would ever penetrate it, I'd promised myself, not even my mother.

I took the next few days off. I called Jan, who was both angry and reassuring; she demanded we go thrifting to take our minds off it. I was so relieved to get out of the flat. But I couldn't shake the image of the orange pickup, or those shielded eyes looking at me.

At home, I constantly checked the locks and looked obsessively through the front window. When I wasn't peeking out, I sat on the couch and stared at the net curtain, terrified at what might be on the other side.

Weeks passed. Maybe he wouldn't return, I told myself, maybe his visit was a one-off. Then, one day, after coming home from work, I was sitting on the couch watching a daytime soap opera. I felt an urge to check the road. I got up and peeked out.

A man was sitting on a motorcycle parked on the side of the road. He was wearing a red helmet, its inky visor facing towards the flat. He lifted the visor. It was my father.

My mouth went dry and my legs felt weak. I ran to the phone and dialled the number of the woman who Mum was caring for.

"He's back." My voice trembled. "He's on a motorbike."

"Oh, bloody hell. Stay put. Make sure the door's locked. I'm calling the police. Is he still on the bike? Is he coming to the door?" I could hear panic rising in her voice.

"I don't know. I'll check." I put the handset down and crept back to the window. The road was empty. A ripple of relief washed over me. I picked up the handset.

"No, he's gone," I said breathlessly.

"I'm still going to call the police. Don't go anywhere. I'll try to get home early," she said.

I sat on the edge of the couch, waiting, getting up to check the road from time to time. An hour or so, later I heard her key in the door.

"I've told the police," she said as she walked in. She put her handbag on the side table. "They told me I can apply for a restraining order. It might take a few days or weeks. I have to go to the Magistrates Court. But, if I get it, he won't be able to come near us legally."

"Okay." I felt slightly more hopeful. "Maybe that'll stop him. He always calmed down when the cops came to Brighton."

Over the following weeks, however, the stalking escalated. Sometimes, he appeared twice in one day when Mum was home. I saw him arrive at random times of the day, the motorbikes and cars were always different. He always pulled up, stared at the flat for a few minutes then left. He never got off the bike or out of the car, and never came to the door.

I didn't go to work. I avoided leaving the flat unless I had to. If I did go out, I scanned the street beforehand. A slithering dread had taken residence in the pit of my stomach. Mum also seemed rattled; she changed the locks. She started peering out the window and double-checking the doors. I imagined him smashing the window and coming in. I asked her about the restraining order. She told me it was being processed.

When he started wearing disguises, my anxiety reached new levels. He turned up in hats, glasses and fake beards. I once saw him with a large, black handlebar moustache. He looked like a macabre version of Peter Sellers in the Pink Panther movies. But there was nothing funny about this.

Finally, my mother got the restraining order. She told me my father had been served with it and warned to stay away. But it would only stop him approaching us; the police couldn't prevent him from being on a public road. The news didn't comfort me.

Mum's mood darkened, her temper became short. We snapped at each other, bickering about housework. We argued about her leaving the flat. I didn't want her to go. She told me someone still had to work and pay the bills.

A week or so later he showed up in a car, wearing a straw hat and dark glasses. I was overcome with hopelessness. If the restraining order didn't deter him, nothing would.

When my mother arrived home, she led me to the couch.

"I talked to the police, Mel. They suggested we should separate. I think it's wise, too. I think he's spying on me, not you. He's obsessed with me. I feel like I'm being followed. I need to move out. You'll be safer without me here."

"What?" I felt the familiar panic. "What do you mean?

Where are you going? What am I supposed to do? You can't just leave me. I can't afford this flat by myself."

I wasn't hearing this. Surely not. Mum went on, quite calmly.

"Sid has offered for me to move in. I think it's the best thing. He's got good security there. I don't think your father knows who he is, and I don't think he knows where he lives. Maybe your friend Jan can move in. You'd like that? Or maybe you can get a flatmate, someone your own age? I'll help you out with the rent until we sort something out." She flashed me a feeble smile.

Anguish turned to anger. I felt my fists clench, my jaw tighten. I had an overwhelming urge to slap the smile off her face. I tried to push away the fury; an inner voice urged me to calm down. But resentment simmered like molten lava. Was she making this up, just to go and live with her new friend? Did the police really suggest that we split up? Was this an excuse to move in with Sid?

"So, you're going to live with Sid." My voice was shrill, like it was coming from a stranger. "So, the police actually told you to do this, did they?"

My mother raised her eyebrows. "Yes, they told me it's safer for us, Mel. Do you think I'm lying? Call them yourself, if you want." Her voice became clipped, her eyes slits.

"Whatever, Mum," I snapped. Acid rose in my throat. "If you want to go, then just bloody go. When are you leaving?"

She sighed. "Well, the sooner the better, really. Sid said I can come straight away. But I'll go tomorrow. I'll just take my clothes. You can keep the cooking stuff, the linen, and other bits and pieces."

"Well, you'd better start packing then." The sarcasm dripped off my tongue like poison.

"Look, Mel, don't be upset. I'm doing this to keep us both safe." Now she was crooning, grabbing my hand. "I'll come and visit. Sid's got a washer-dryer, so you won't need to visit the launderette anymore. I'll collect your washing and bring it back for you. I'll give you Sid's number, so I'll be here in a flash. I'm only fifteen minutes away."

She was trying to sound gentle, soothing, caring and supportive. My face was burning. "What about Dad? What if he keeps coming?"

"I doubt he will, once he realises I'm not here. He's not interested in you. It's me he's stalking. It'll be okay, Mel. Even if he shows up, he's not allowed to approach you. If you see him, call the police. But I don't think he'll come."

She smiled. "I'll invite you over for tea one night, so you can meet Sid."

The following day, I watched my mother pack her car. I wondered if Dad was also watching from the shadows.

As I stood in the doorway, looking through the ripped flyscreen, I felt like a little girl again, watching my mother leave me. This time, however, I wouldn't throw myself at the car. I wouldn't plead with her to stay. I wouldn't cry. I swallowed my sadness as she got in and drove away.

I was on my own.

LATER THAT EVENING, I called Jan.

"She just left you?" I heard the concern in her voice. "You're there all by yourself?"

"I've got Blitz," I replied. "And Mark said he'd pop around when he has time."

"Who is this Sid guy? Is that the one she met through the paper?"

"Yep. I've never met him. Mum says he wants to help her. Apparently, he lives in a posh place at Seaton. He's Sri Lankan, like Dad."

"What is it with your mother and Sri Lankans?"

That made me laugh. Then she stunned me.

"Look, I'm over school. I don't want to finish Grade 12. How about I do something radical and move in for a while? I'll get a job. It'll be fun."

My heart leaped. I was deeply touched. I told her to come whenever she wanted.

A week or so later, Jan and I became flatmates. I gave her Mum's key. Having company felt good. And since Mum had left, I hadn't seen my father.

Jan told me she'd got a job at a supermarket in the city; they'd promised her full-time employment if it worked out. I picked up more shifts at the restaurant.

We started going out again, coming home in the early hours of the morning after partying at our favourite hotel, stopping for a breakfast of pizza on the way home. We were drinking more and experimented with weed. I didn't like it. The feeling of not being in control scared me. We even tried a stint as late-night go-go dancers in a local nightclub. That abruptly ended when the owner demanded we go topless.

I met a young guy whose parents had recently purchased the hotel. His name was Kev. He'd moved to Adelaide with his family from Victoria. Sometimes, he worked behind the bar and gave us free drinks. I liked him. He asked me out. I declined, saying I'd rather be friends. He grudgingly accepted, on the condition we could hang out together. I agreed.

Mum visited occasionally, mainly to pick up my washing or drop it back, dry and fluffy.

I asked if she'd seen Dad. She told me she hadn't, but still felt like someone was following her. I told her it was probably her imagination, although secretly, I was worried he was monitoring her movements and planning something nasty. I didn't trust him.

I forced myself to stop checking the window so regularly. I tried hard to put my father out of my mind. Instead, I focused on earning enough cash to splurge on Friday and Saturday nights.

One evening, my mother invited me over for dinner to meet Sid.

It only took me ten minutes to get there. Light was fading as I pulled into a large driveway laid with grey pavers in a herring-bone pattern. Three identical terracotta townhouses sat side by side. They looked like dolls' houses. Each building had a small square patch of green, dotted with elegant white and red roses. For a moment, I wondered if I'd accidentally driven the Mini down a rabbit hole and was about to take tea with the Mad Hatter.

Finding the right house, I rang the bell.

A tall, balding man opened the door. Thin wisps of black hair lay flat against his shiny scalp. He was paunchy with dark acorn skin and midnight-blue eyes, surrounded by deep rivers of wrinkles.

"You must be Melody," he said warmly. "Come in, come in." I heard a slight Indian accent. He took my hand and shook it. "I'm Sid," he smiled, cheeks dimpling.

I looked at his hand in mine. It didn't seem to fit the rest of him. His fingers were long, narrow, bony and delicate. Later, I found out he'd played the piano and had owned one before retiring.

I followed him in, gasping as I crossed the threshold. The air smelled like freshly cleaned leather, like a new saddle. The plush cream carpet looked luxurious. I resisted the urge to roll in it.

"I should take off my shoes," I said. "I don't want to ruin your carpet."

"Yes, yes, of course, my dear. Take them off, leave them here at the door."

I reached down and removed my sneakers. Barefoot, I wiggled my toes, basking in deep soft pile.

"Is that Mel?" I heard my mother's voice from somewhere inside.

"Yes, darling," he called back.

I was surprised. While I'd suspected they were more than friends, this endearment confirmed it. I'd never heard anyone call my mother by a schmoozy nickname before. It made me feel a bit uncomfortable.

As if sensing my unease, Sid led me to a lounge room with a plump chocolate leather couch. I sank into it. It was the most comfortable thing I'd ever placed my bottom on.

"I understand from your mother that you drink occasionally," he said. "I don't condone underage alcohol consumption, but I'm happy to offer you a glass of white wine."

"Only if you're having one," I replied a little hesitantly.

"No, no, no." He shook his bony finger at me and chuckled. "I'm diabetic. I also have heart issues and high cholesterol. No wine for me. I'll stick to mineral water."

"Darling, will you pour your daughter a glass of wine, please?" he called to a doorway at the rear of the lounge. I assumed it was the kitchen.

"Be there in a tick," she sang back. I couldn't recall ever hearing my mother sound so cheerful.

I felt like I'd walked through the wardrobe and ended up in some parallel universe. Since when had my mother been so agreeable? How was it that, only a short time ago, we'd been peering fearfully through the curtains to see if Dad was outside? Who was this strange, lanky-handed man with his squishy couch, which looked straight out of a lifestyle magazine? I was expecting Mr. Tumnus to pop out of the kitchen at any minute, carrying crumpets.

Instead, my mother appeared. Her hair was perfectly curled, lips tinted with a hint of pink. She was wearing a frilly apron. I felt my jaw drop. An apron! I wasn't in Narnia. I was in a 1950s sitcom!

"Hi, Mel," she beamed. "Here's your wine. Do you like the townhouse?"

I took the glass, craned my neck and looked around. The décor was tastefully neutral. A wide, carpeted staircase led upstairs. I'd never been in a house with two floors before.

"We even have a microwave." She smiled. "Sid thinks microwave cooking is healthy. He made tonight's soup in it."

Sid sat next to me. I could hear the whoosh of air as he eased into it, like the couch was decompressing.

"You have a very nice home," I said.

"Well, I've worked for it. You don't get anything in this life without working for it. Now, I can enjoy the fruits of my labour." He smiled. "I like dusting, keeping things clean. I have a place for everything, and everything in its place. I'm teaching your mother how to cook." He turned to her and winked.

She was beaming. She looked ten years younger. I couldn't take my eyes off the apron. It had small white flowers on the frills.

A mixture of emotions stirred. I was happy for her, but also strangely envious. Part of me wanted to get up, run around and knock the perfectly placed ornaments off the shiny shelves. I fought an urge to pour my wine on the beautiful carpet. I felt jealousy gnawing at my stomach. I pushed it aside, telling myself to behave. I was a guest. I had manners.

Sid asked me what I was going to do about my education. I told him I didn't know. Maybe I'd find a full-time waitressing job. He snorted and said I shouldn't do manual work all my life.

We sat at a gleaming, oval table of burnished mahogany and ate with matching cutlery. I found it hard to swallow, the taste of envy overwhelming my senses.

"Your mother is a beautiful woman," Sid told me, as he ladled a second serving of mushroom and spinach soup from a large tureen into small matching orange bowls.

"I'll be happier when the divorce comes through, though. Sounds like your father is a real psycho. She's better off cutting all ties. That'd make me feel much better about this whole situation."

After dinner and tea sipped out of dainty porcelain cups, I left carrying containers of leftover soup and chicken curry. I felt like Alice in Wonderland, laden with goodies, going back up the rabbit hole to reality.

I drove home. When I walked into the flat, it looked shabbier than usual. I greeted Blitz, put the leftovers in the fridge and plonked myself next to Jan on our ratty couch.

I realised I didn't have much, but I did have my independence. Mum could swan around in her luxurious townhouse with pernickety Sid.

I told myself I didn't care. But I did.

10

THE ATTACK

ONE MORNING WE change the routine. We're on the water early and Mike is keen to visit the other dolphins while it is still calm and swing by to see Jock on the way home. Steve is also with us that day.

After documenting six or eight dolphins, some of them well-known regulars, we turn into Jock's patch.

As usual he is circling his boat. When he sees us, he breaks off, issues a loud raspberry and follows us into our private playground.

As he trails behind us, we notice something hanging off his dorsal fin. It's a mass of fishing line and hooks that have become entangled in his already mangled flesh.

"We have to get that off," says Mike. "Otherwise it's going to cut into him and probably catch on more discarded line."

Poor Jock, I think, you're always getting yourself into trouble. I feel a surge of empathy. We have to do something and it has to happen now.

"Okay," I say. "How about Steve and I jump in. We will lure him to the shallower end of the channel. That way we'll have a footing. Hopefully he won't obsess about getting our feet off the bottom. Maybe Steve can grab him and hold him while I try to get the line off. Do we have anything to cut it off with?"

Mike quickly retrieves a pair of scissors from a box. "You have to be careful, mind your hands. I saw hooks wrapped up in that line and they're sharp, and they could be rusty. Have you had your tetanus?"

I ignore the question and take the scissors.

Steve and I slide into the water. Jock immediately approaches us. Then, he stops. It's as if he suddenly realises we have scissors and a plan and all at once he becomes wary. He hangs back, darting every time we make a move towards him. I lunge towards him, scissors in hand, hoping somehow to grab and cut all at once. He evades me easily. He seems to know we are up to something.

This game of cat and mouse goes on for more than ten minutes.

Steve inches up behind him and in a swift movement wraps both arms firmly around Jock's torso. Jock starts thrashing in his grasp, his strong tail fluke slapping and flapping. I can see Steve's muscles bulging with the effort.

"We've got to keep his blowhole above the water," I yell at Steve, salt stinging my eyes.

Scissors in one hand, I approach Jock's fin. Water everywhere, spraying, splattering. I can barely see.

"Just do it!" yells Steve. "For God's sake, Mel, I can't hold on!" I eye the spot where the fishing line has wedged in a fold of skin. I cut. I yank. I feel something sharp pierce my own skin. A bundle of fishing line with half a dozen small hooks lies in my hand. We've done it.

Jock instantly swims away. He doesn't return to us that day. Exhausted, we heave ourselves back onto the boat. I hope he still trusts us. But I'm so relieved we've managed to help him.

A WEEK LATER, *we head out again, the fishing line incident with Jock still fresh in my mind. I worry that due to the strange shape of his dorsal, more entanglements are likely. But today something else is troubling me.*

The pewter sky is rapidly turning a deep charcoal. Storm clouds are engorging.

Although the wind isn't picking up yet, I know it's coming. We only have a short window of time.

The weather gods must be having a sick joke at our expense, increasing the sense of impending doom. I don't want to take this trip.

Mike had received word from a fisheries officer that a dead dolphin had been found on one of the riverbanks. Apparently, it had a hole in its body. They'd planned to collect it, but Mike wanted to ensure it wouldn't get washed away with the tide before they did. We set off without speaking.

It's not long before we find the location. We drop anchor. Mike wades towards the bank. I watch from the boat. It reminds me of the time Mike had told me about how Jock had been attacked with a flounder spear. He'd survived but the spear left scars on his torso. I was dumbfounded by human cruelty. Why would anyone attack a dolphin?

"Yep," he calls back. "It's here. It's only a young dolphin, so they should be able to collect it. It's lodged in the mangrove roots, so it shouldn't move. Let's stay here until the officers come."

He turns and stumbles back. From my vantage point, I can make out a sliver of grey, motionless in the mangroves' embrace.

"Did you see any marks?" I ask, as he clambers back into the boat.

"From what I could make out, it's got a hole in its side. Maybe a spear hole, it could even be a bullet hole."

He's angry. I think I'm going to throw up. Why would anyone attack a dolphin so brutally? We sip coffee and wait, hearing a faraway rumble of thunder. The officers turn up in their inflatable, load the body onto a sling and bring it back to the ramp. It will be sent to the museum for a post-mortem.

We stand on the ramp, surveying the body. It's only a small dolphin so probably quite young. In its side, there is a hole about the size of a dollar coin. Two thin red lines snake down from the wound to its belly.

"Well, it's either a spear gun or shotgun, by the looks of it," Mike says to the officer. "Hard to tell. Only a juvenile though, and I don't recognise her. She doesn't have any notches in her fin."

I walk to the nearby washroom and vomit. After heaving up the coffee, I splash water on my face with trembling hands. This doesn't seem possible. How can a human be so inhumane? Had this dolphin tried to steal someone's fish, which triggered a retaliation, or had someone killed this animal just for fun?

I return to the ramp. The officer is helping to put the body in the back of a van.

"It's just so bloody senseless," Mike mumbles. "Just so senseless."

Later, feeling more composed, I reflect that I've never witnessed a dolphin kill anything which it didn't intend to eat. Why is it that humans do? Does this go back to our disconnection with nature? Our desire to rule over the natural world, to control it, manipulate it and use it instead of being part of it? Where is the compassion, the respect for other living things, particularly those that bring so much joy into the world? Why anyone would want to snuff that out is beyond me.

Perhaps we all get lost in the dark sometimes. I understand that well. But I've come to realise in my adult life that nature, and dolphins in particular, can inspire us to find the light. Their exuberance for life is infectious. For them, there appears to be no such thing as premeditated violence and hatred. To me, they seem incapable of such viciousness.

As a human, however, I know and accept that violence will always be a part of my world. Even if I watch it from afar. I have to find a way not to be consumed by it, to focus on kindness and goodness and believe in justice.

LIFE WAS FINALLY falling into a comfortable rhythm.

Jan had been living with me for a few months. I loved having her around. My father hadn't returned.

One afternoon in April 1986, after a busy lunch shift at the restaurant, I decided to curl up on the couch with a pack of Yo-Yos and a book. It was a beautiful day, with filtered sunlight streaming through the net curtains.

There was a knock at the door. Slightly annoyed at being disturbed, I heaved myself off the couch. My mother was standing outside, holding a basket of washing.

"Hi, Mum."

"Hi, Mel, here's your sheets. I used a softener, so they should be nice and snuggly."

I thought I saw a movement out of the corner of my eye. I dismissed it. I looked down to wedge open the screen door.

Out of nowhere, my father appeared behind her.

"Mum! Watch out!"

"What the—?" She tilted her head to one side, brow furrowed, eyes wide.

My father lunged at her.

The basket fell, the fluffy white sheets splayed across the asphalt. He wrapped his arm around Mum's throat, gripping and pulling her to him. I could see the muscles in his forearm bulging as she struggled.

A scream ripped through my throat. My mother started screeching, high-pitched and intense.

My legs were glued to the floor. Time seemed to stop. The scene in front of me wasn't real.

I noticed something in my father's hand. He slapped it to her face. He was pushing, grinding the thing into her.

My mother gave a piercing, agonising wail. It was blood-curdling, non-human.

"Mel, help me!"

My father glanced at me with a malicious grin. He let her go, shoving her forward. She collapsed across the doorway, landing on her side.

My legs became unstuck. I jumped back, looking down. A horrific realisation. Something black was poking out of her cheek.

I clutched at her top and dragged her inside. She was howling, eyes closed, thrashing wildly on the floor. Blood covered her face.

"Get it out!" she screamed. "Mel, please get it out." Her anguish was unbearable.

My mind whirled. The room tilted. "Mum, towels, I need towels." I staggered to the bathroom and grabbed a towel from the rail. I braced myself against the wall.

Don't you faint, Mel, I told myself. I stumbled back and dropped to the floor beside my mum.

"Get it out of me!" Her eyes were tight shut, her face twisting and turning like a fish gasping for air.

"You've gotta stay still, Mum." I grabbed the thing protruding from under her right eye. I recognised it. It was the corkscrew Mum used every year to open her Christmas sherry at Brighton.

The handle felt slippery, slick with blood. I pulled gently. It didn't move. Mum's screaming got worse. My vision blurred with sweat. I took a deep breath and inched the handle anti-clockwise. It gave a little. I turned it slowly. My hands were sticky and wet. I felt the grind of metal on bone. The screw started emerging from her flesh.

I kept turning gradually, transfixed on the task. My mother's screaming was coming from somewhere distant. With one last gentle twist, it was out. I threw it across the room.

"I got it, Mum. It's out!" Suddenly, I felt on the verge of laughing hysterically. I looked at her face. Blood was oozing out of the hole. I grabbed the towel, bundled it on top of the puncture and pushed down. She thrashed, howling.

"I've gotta call an ambulance. Hold it firm to your face," I ordered.

She raised her hand to the towel. I got to the phone and, with quivering hands, dialled the emergency line.

"My mum, she's been stabbed," I said. "I need an ambulance and the police." I gave the address.

A terrifying thought struck me. I hadn't locked the door. I raced to the entrance. The flyscreen had swung shut. I stepped up to it, opened it a crack and peered out. My father was on his back, in an expanding sea of red. A carving knife was in his right hand. Deep gashes criss-crossed his throat. His eyes were closed. I thought he was dead.

I ran back inside, locking the door behind me. I called the police again, then returned to Mum. I lifted the towel. The bleeding had slowed. I could see a deep hole in her cheek-bone, a couple of centimetres below her right eye. She was sobbing.

"It's okay, Mum," I said. "Help's coming." I stroked her sodden hair. The brown, ruddy streaks felt stiff under my hand. "I think Dad's dead."

She didn't seem to hear me. Her mournful sobbing continued.

The sound of sirens. I went to the door and escorted the paramedics inside. One of them wrapped me in a blanket and told me to move aside. The other went to my mother and started talking to her softly.

They examined her and spoke into walkie-talkies. A short time later, I heard more sirens. Paramedics wheeled in a stretcher and loaded Mum onto it.

There was a commotion outside. Through the window, I saw uniforms and another stretcher. Something inside of me broke apart. I started weeping uncontrollably.

I don't know how long I stood in the corner, watching, shaking and crying. The enormity of what had happened was dawning on me. Then I felt an arm circle my shoulder. I looked up. It was a man in a suit with a badge on his pocket.

"Melody, it's going to be alright," the man's deep voice said gently. "My name's Detective Dunstone. You can call me Peter." His eyes were full of sympathy.

"Your mother's going to be okay," he said. "She could've been blinded. The EMTs think the corkscrew missed the nerves to her eye."

"Is Dad dead?" I asked.

"He's on his way to the hospital. I don't know if he's going to make it. He's lost a lot of blood. You probably saved his life. A few more minutes, and he'd have been gone."

I felt numb.

"I'm going to Flinders Medical, that's where they're taking your parents. Want to ride with me?"

I nodded.

"Keep your head down outside, okay? A few TV news crews have turned up."

He pulled the blanket over my head, kept his arm around me, pulled me close and ushered me outside. From inside the blanket, I spied one-eyed monsters sitting on men's shoulders, following our movements.

We got into an unmarked police car and drove to the hospital.

DETECTIVE PETER DUNSTONE was right. The corkscrew hadn't hit any major nerves, but my mother would live with a hole in her cheekbone and a deep scar—a constant reminder of the terrible attack. She was released the following day.

According to the doctors, my father had been close to death due to severe blood loss. They confirmed that, if my call to the paramedics had been any later, the outcome would have been different. He was staying in intensive care.

Peter told me it would probably be weeks before my father would be fit enough to face charges.

I had mixed feelings about my part in saving my father. I didn't want to be responsible for anyone's death, but later, I did wonder if I had known, or had my time again, if I would have responded differently.

Later that evening, Peter dropped me back to the flat. Jan was waiting there for me. I had called both her and my brother from the hospital.

"Oh, Mel. God, you poor thing." She hugged me fiercely as I walked in the door. "The cops have been here all afternoon, taking pictures and dusting things. I gave them a wide berth. Are you okay?"

She looked so concerned.

"Yeah, I'm okay," I sighed, suddenly feeling exhausted. "I had to pull that thing from her face, Jan. I don't know how I did it."

"I'll get you a drink." She tried to smile. "I think we have some of the hard stuff left."

The lounge smelled like disinfectant. It reminded me of the hospital. I guessed Jan had cleaned up the blood on the floor. She walked back in, holding two glass tumblers filled almost to the top with amber fluid.

"Here," she said, handing me one. "This should make you feel a bit better."

"Did you mop up?" I asked, taking a sip.

"Yep, it's not perfect but it's better. Outside is a bit of a mess. You probably didn't see it coming in, but just don't look tomorrow. It needs a hose off. A guy came around and said he'd do it. I think he owns this place."

I sank back into the couch. Jan sat next to me. We sipped in silence.

"I can't stay here, Jan," I said. "There's just been too much shit."

"Yeah, I figured as much," she replied. "I called my mum earlier and we talked. I'm going to move back home for a while. I'll wait until you find another place though."

"Thanks," I said, placing my head on her shoulder. I felt so weary.

"Do you think you can move in with your mum and her boyfriend?" she asked.

I turned and smiled. "I don't reckon they'll ask. I think three's a crowd and, besides, I'd mess up his immaculate home."

Jan giggled.

"Are you going back to school?" I asked.

"No. I really like my job. They think I could eventually be management material, so I think I'll stick it out."

We finished our drinks and I collapsed into bed. Peter had given me his business card, telling me I could call him any time. I dozed off, thankful I had someone I could rely on.

The following day, the man hosed off the front of the flat. He was the caretaker, he said, the owner's son. I told him I'd be cutting out of the lease. He seemed to understand.

Jan had taken the day off work and was helping me to give the flat a thorough clean. I wanted no trace left of the previous day's events.

Mark phoned. He was a fully registered nurse now. He was still living at a hospital in the city, but not the one my parents were taken to. He told me he'd visited Mum and she was doing well. She was now at home but Dad was still in intensive care.

As I relayed what happened, we both cried.

"I don't think Dad meant to kill Mum," he said, sniffling through the earpiece. "I think he wanted to blind her, or maybe mutilate her face to make her unattractive."

I considered it. It seemed plausible. If my father had wanted to kill her, surely, he would have used the knife he had cut his own throat with? I told Mark I didn't want to think about it and that I needed to leave the flat.

"Mel, I get it. You've got to go. You can't stay there," he said. "It's about time I left the residential wing here, so how about we look for a place to share for a while? I don't want a flat. I've been cooped up in this small space for too long. Surely, with us both working, we can get a small house or something."

A warm feeling spread through me. My big brother was coming to the rescue.

"Okay," I said. "I'd really like that. Thanks, Mark. I love you."

A few weeks later, Mark and I moved into a house just a stone's throw from Sid's place. It was cream brick with brown roof tiles. It had two large bedrooms, a kitchen with dark timber cupboards and enough space for a small table, and a separate lounge. The backyard was compact, but had a patch of lawn and two peach trees. Blitz was happy to be outside again.

Jan moved back home, promising to visit as often as she could. Mark and I went shopping for second-hand furniture. We found a couple of faded red two-seater couches, a humble pine dining table and chairs, two double beds without headboards, and some chipped end tables. He brought his TV from his room at the hospital. We bought new sheets and towels, and a big oak cupboard that we would share.

Mum came to visit. She wore a patch of white gauze on her cheek, secured with tape. She said it was painful. She refused to talk about what had happened at the flat, insisting that she needed to put it behind her. We'd also been told by Detective Peter Dunstone that we shouldn't share stories, because my father was likely to be charged and we might be needed as witnesses.

My friend Kev from the hotel came to visit regularly. I always heard him before I saw him—the loud rumble of his bottle-green V8 Holden pickup was unmistakable. He always brought beer for Mark and sparkling wine for me. We sat in the backyard on our wonky kitchen chairs, drinking and talking, with Blitz chatting away in the background. Jan sometimes joined us.

I told Kev about everything—growing up, leaving Brighton and what had happened at the flat. He told me he couldn't really comprehend it. His family was close.

Kev was kind and funny and three years older than I was. I liked his roguish blue eyes and ready smile. He hiccupped when he laughed. He was chubby, with receding mousebrown hair and a long nose. He wore moccasins, which I thought were ugly. I thought that, maybe, I would go out with him when I felt ready. I felt safe in this new house. My father couldn't get to me now. But he'd get out of the hospital eventually, and I dreaded that day. I wasn't ready to face another ordeal.

11

TESTIFYING
AGAINST
MY FATHER

THE MORNING IS CRISP. *The yawning beach looks flawless, unnatural, Photoshop-enhanced. The water lapping at the shore is implausibly perfect, liquid silica infused with a hint of azure. The scene makes my eyes hurt. It's almost too beautiful to look at.*

It's late 1992, and we're at Monkey Mia, 900 kilometres north of Perth, Western Australia, where wild dolphins routinely swim in the shallow waters to be fed by people. I'm visiting the tourist attraction with Mike and fellow research assistant Steve to learn more about these dolphins and the dynamic of their human interactions.

Along the shoreline, a dozen people are looking out to sea, wearing shorts or rolled-up jeans. Other people stand in the water facing the beach, holding buckets of fish.

The dolphins appear. One by one, people on the shore step forward until they're calf-deep in water. The bucket bearers hand them each a dead fish. The visitors bend down and the dolphins wait, bodies tilted at odd angles, eyes staring upwards at the offerings.

More fins approach, navigating the shallows, turning on approach, contorting their bodies to fit in the tight space between the sand and water's surface.

The air fills with a chorus of excited squeals, laughs and cries. We trudge down the sand to the surf and watch from behind. Then a young woman with a bucket smiles at us.

"I'll hand you a fish," she says. "And you gently bend down and give it to the dolphin. They won't bite, just be gentle and respectful. After feeding, please step back."

The whole activity seems so contrived and transactional, but I step farther into the water and take a fish. A dolphin approaches, inching its way towards me sideways, shimmying like a belly dancer. It opens its mouth wide, displaying rows of sharp teeth and a salmon-pink tongue. It looks like it's begging.

I do as instructed, dropping the fish into the gaping mouth, then turn to walk away. The deal is over. I watch the dolphin wiggling and shuffling back to open water. I can only think of Jock—how lucky I am to have a relationship with a wild dolphin that doesn't require any coercion, any tasty enticements. Our interactions are always on his terms, but I like to imagine we're somehow communicating and enriching each other's lives.

But part of me wonders if in some way, Jock has become too reliant on us and Mike's other research assistants for companionship. Again, niggling worry gnaws at my insides. I had come to love, even rely on Jock's friendship, but had he become too dependent on us? Were we doing the right thing by him?

The contact with the dolphins at Monkey Mia seems artificial, forced and superficial. I understand the desire to get up close and personal with a wild dolphin, to experience that apparent bond, for a moment. I also know that the program helps teach others about wild dolphins—and that is valuable. Yet I hate seeing these wild dolphins being—in my mind—little more

*than trained performing animals. I have trouble with the idea of
people capitalising on one of dolphins' basic needs just for one
fleeting moment.*

*Mike says he's spoken to other researchers who have con-
cerns about the mother dolphins becoming so dependent on
the fish handouts, they aren't teaching their youngsters how to
fish. These mother dolphins aren't abandoning their duties on
purpose—they've just become accustomed to having human
handouts to survive.*

SOON AFTER VISITING *Monkey Mia, I book a short trip to
Queensland in my mid-semester break and include a trip to a
well-known marine theme park. I've phoned ahead, explaining
I'm helping with research on Adelaide's dolphins and am kindly
offered a behind-the-scenes tour. I've learned a lot about cap-
tive dolphins from Mike but feel I need to see them for myself.
As I disembark from the plane, the steaminess of Queensland in
October slaps me in the face and leaves me breathless.*

*I feel nervous. I know I have to keep my emotions in check
and maintain an open mind. I know the park has done a lot of
good work, rescuing whales entangled in fishing line and nets
off the coast.*

*At the gate, I'm met by a smiling young man called Todd. He
tells me it's great to meet another dolphin-lover. Todd explains
that the next dolphin show is an hour away so that will give us
time for a tour.*

*He leads me across a big empty stadium. Hundreds of seats
hug a large sparkling pool. There are a couple of dolphins in the
pool, swimming around the edge. Unlike the scarred and nicked
Port River dolphins, their dorsals are smooth and perfect.*

*A young woman with a whistle on a lanyard is kneeling, one
hand in a plastic bucket. I've arrived just before the morning feed.*

The dolphins swim to her and accept a snack. She blows the whistle, claps and speaks to them like they're kids in a kindergarten.

Todd leads me to the back, where visitors are restricted. Apart from a faint whiff of fish, combined with the tinge of disinfectant, it looks as spotlessly clean as any hospital ward. Todd talks about all the rescue work the park has done over the past year, and the importance of animal welfare to the staff. There's little doubt that he loves the dolphins in his care.

I walk around the pools, where dolphins who aren't performing are sometimes kept and where any sick animals are treated. I tell Todd politely that, while everything looks clean and sparkling, I'm not a fan of dolphins in captivity.

During the show, I feel as if I've been transported to an alternative reality. Beaming trainers enthusiastically encourage perfect smiling dolphins to leap through hoops and tail-walk. I think back to Billie, who seemed to enjoy tail-walking. I muse about how different this is to the Port River. While Billie had learned the manoeuvre in captivity, she chose to continue it in the wild. Here, the dolphins don't have that luxury. Performing is their job. They're paid in fish. The crowd cheers and applauds every time one of them balances a ball or back-flips through the air.

I leave before the show finishes. The sights and sounds are too much for me. I find the whole thing disturbing. I understand the park is, in its own way, raising awareness about dolphins and people flock to see them there. But, while physically perfect, these animals are shadows of the dolphins I've come to know in the Port River. I wonder if, given the choice, this would be the life they'd want. I feel sorry for them. For all its flaws, the Port River is a wondrous place, a nursery for young fish, an ecosystem supporting all kinds of life. My dear friend Jock and the other dolphins there face many dangers, and many are scarred from their experiences. I also have the ever-nagging concerns about

Jock's reliance on us. I love Jock but I'm becoming more aware that his friendship with us and implicit trust are making him more vulnerable to humans who may want to exploit or harm him. I know, too, that in a way, Jock is imprisoned, unable or unwilling to leave his part of the river. His cage seems to be self-imposed even though he is free.

The dolphins at this park have no choice. I wonder how they cope, having to exist their entire lives in an artificial environment, bound by human-created rules.

MY TRIP TO QUEENSLAND *was a shock to the system and brought me back to my own trials. In a strange way, it caused me to drag up a moment in my life when I had felt trapped, unable to escape a man-made legal world filled with rules and conventions, to which I had never before been exposed. I recalled the panic and fear I felt, when I was asked to behave and even perform in a certain way in front of a crowd, on an unfamiliar and intimidating stage. That stage was a courtroom. The words I uttered there would help determine the fate of my father.*

IT FELT LIKE MONTHS, but it may have only been weeks. I was approaching eighteen years old.

Kev and I slowly grew closer, and our relationship blossomed into romance. I wasn't sleeping well. The familiar routine of waking with a thudding heart and an urge to flee had returned. My dreams were filled with shadows.

Peter Dunstone, the detective, checked in often, asking how I was doing and if I needed anything. I was touched and surprised by his concern. He seemed rock-solid—dependable and caring. I told him about my night terrors, the flashbacks to what had happened at the flat. Everything would be okay, he assured me. For some reason, I believed him. He suggested that, when I felt ready, perhaps I should consider getting some counselling. I took up this suggestion some time later.

One day, he called with news about my father.

"He's been released from the hospital," he said. "We've charged him. He's been remanded in custody."

"What does that mean?"

"It means, the case will go to trial in the Supreme Court. A jury will decide his fate. Mel, your mother will be testifying. We need you to as well, if you can. I know it will be hard, but I'll be with you all the way."

I felt a tightness in my chest. "I don't mind testifying, if it's going to help, but how can he deny it? We don't actually need to prove it, do we? I was there, I saw what happened."

"I wish it were that simple, Mel. Your father says he didn't attack your mother deliberately. He claims it was an accident."

I was speechless. I drew in a breath. "It was no accident."

"I know, I know. Listen, we should have a court date soon. The Crown Prosecutor, the guy who's going to try to get justice for you and your mother, his name is Peter Brebner. He's a good guy and a great lawyer. Stay strong. I'll be in touch soon."

I hung up, feeling shaken.

During the following few days, apprehension hung over me like a thick, stifling cloak. Even when I was outside, I felt claustrophobic. I couldn't shake it off. I worried that my father would use his intelligence and charisma to convince the jury he was innocent. My mind was spinning with questions. What if he gets off? What if it starts all over again? Will he leave us alone this time? He'll want revenge. What if I mess it up and they don't believe me?

Mark and Kev both tried to assure me justice would prevail, that I should trust in the system. I wasn't convinced.

I didn't share my concerns with my mother. When I visited her, she was obsessed with the hole in her face. I consoled her as much as I could. I knew she was in pain and might be disfigured forever. I told her it would only leave a small mark, barely noticeable. She'd still be just as beautiful. She never asked how I was.

The day of the hearing arrived. It was January 1987, a gloriously sunny summer day, but my eyes felt heavy and sore. I had lain awake the night before, worried I was going to forget something, or say the wrong thing. When I did nod off, the night terrors returned. I didn't feel ready.

We were to meet Detective Peter Dunstone in the foyer of the Supreme Court. As Mum and I pulled up in the taxi, I was struck by the court building's majesty. It reminded me of a Roman palace. Its imposing honey-coloured stone was embellished with huge columns. Elegant carved scrolls sat above arched windows. The building exuded strength, authority and power.

I felt nauseous as we walked into the cavernous foyer. I was conscious of my kitten heels clicking on the marble floor. I was wearing a simple brown dress, the pockets stuffed with tissues, and my good black dress shoes.

My mother no longer had the bandage on her cheek. The skin was puckering where the stiches had been. She seemed composed and in control. She said Sid planned to meet us for lunch, but Mark wouldn't be coming. He'd said he had to work.

Peter Dunstone was waiting. He waved and approached us. He shook my mother's hand and put an arm briefly around my shoulders. He was wearing a tie. I'd never seen him wear one before. I thought he looked smart and professional. I wondered how old he was, guessing he was in his thirties. He smiled at me warmly. I managed to smile back, relieved he was there.

"We're going to wait in a room until you're called," he said. "Follow me up."

He led us to an elevator, which took us to a large open space with multiple doors opening off it. Each door was marked with the words "Meeting room" and a number. There were also bigger doors with the word "Courtroom" and a number.

Other people were walking around, talking in hushed tones. Some wore long, black flowing robes and tightly curled wigs. I found it odd that I could still see their hair under the wigs. What was the point of wigs, I wondered, if they didn't cover your whole head?

Peter led us to one of the meeting rooms. It had plain, cream walls and was sparsely furnished. A cranberry-coloured vinyl couch sat against a wall. It looked stiff and unwelcoming. Opposite the couch, two matching armless chairs stood at attention, while a tastefully plain timber table squatted in the centre.

Mum and I sat on the couch. It was as hard and uncomfortable as it looked. Peter closed the door and sat on the edge of a chair. His eyes now looked solemn, intense.

"The prosecutor Peter Brebner will be coming in to see us shortly," he said. "He'll then go into court and give his opening statement, as will your father's lawyer."

"So, we don't go in together?" my mother asked.

"No. We'll be called into court separately. I'll be going in first. I'll stay while each of you come in. You'll be escorted back here when you're done."

A knock at the door. A man wearing robes, but no wig, whooshed into the room. His arms were filled with folders and papers. He looked much older than Peter Dunstone. He wore silver-rimmed glasses.

"Hello, I'm Peter Brebner," he said, with a quick smile. He held out his hand to my mother, then to me. He towered over us.

"I'm sure Peter has run you through what's going to happen today. I'll be asking you some questions. Do you have any for me?"

"Will Neil be in the courtroom?" my mother asked.

"Yes," he replied. "He will be seated in the dock. I recommend that you avoid eye contact with him. Talk to me or his lawyer. Just answer the questions and tell the truth. There will be a jury in there, too. You can look at them if you wish. Don't worry. I'm confident this will all be over soon."

My mother nodded and flicked a glance at me.

"Righto, then. See you in there. A clerk will come and get you when it's time." He turned and swished out.

We sat in silence. I gnawed at my nails, wishing I'd brought a book. My mother kept rifling through her handbag for something she didn't find. Peter Dunstone popped in and out of the room.

An hour or so passed. A clerk came in and told Peter he was required. He winked at me encouragingly as he walked out.

My mother kept checking her watch and clearing her throat. She left, then returned with two plastic cups of water.

About forty minutes crawled by. A knock. The same clerk entered the room and escorted Mum out.

I was left alone, with a churning gut and deepening dread at the thought of seeing my father again. I wasn't sure if I could keep it together. I didn't want to break down in front of him, show him weakness.

THE KNOCK AT THE DOOR jolted me from my dark thoughts. The clerk came in.

"Ms. Horrill, can you please follow me?" she said coolly, her young face impassive. "We've just had a short recess, but we're about to reconvene. If the Judge speaks with you, please refer to him as Your Honour and be respectful."

I nodded, got up from the couch and wiped my hands on my dress.

She opened the door to the courtroom and ushered me in, leading me to an elevated, partially enclosed area against a side wall. I stepped up. Three timber partitions surrounded a maroon high-backed chair. A slim, black microphone jutted from a wide ledge. Next to it sat a nut-brown book with the words "HOLY BIBLE" emblazoned in gold across the cover.

"Remain standing, please," the clerk said.

I did as she asked, although my legs felt rubbery.

The air had a slight musty scent, like inside a house that hadn't been aired for a while. The only discernible noise was a low murmur of voices.

I looked around. Magnificent panels of amber timber covered the entire front wall. A huge carved Australian Coat of Arms presided over the room. A row of claret-coloured chairs nudged against a timber bench, stretching across the room. Some chairs were filled with men and women in black robes and wigs. They were in deep discussion, shifting pieces of paper here and there. Above it, a single large chair sat empty in the middle of another bench. Below, a long timber table held more robed, wigged people seated and in conversation. Its surface was littered with books and folders. I noticed the Crown Prosecutor Peter Brebner. He was holding papers and talking to Peter Dunstone. Neither of them noticed my arrival.

At the far right of the table sat a woman with what looked like a typewriter. I thought she'd have to be an extremely fast typist if she had to record everything said in court.

Behind the lawyers, facing the front, were rows of simple timber bench seats. The set-up reminded me of a movie theatre, but you wouldn't want to sit here for a long time.

Some seats contained people dressed in everyday clothes. I scanned the rows and saw my mother. Our eyes met. She smiled and mouthed, Alright? I nodded.

Out of the corner of my eye, I caught movement. I turned. My father, flanked by two men in uniform, was easing into an elevated chair in a partitioned-off area on the opposite side of the room. Head down, he appeared to be immersed in the task of sitting.

I couldn't take my eyes off him. He looked haggard. He seemed to have aged years. His once-handsome face now looked gaunt and grey. Peppery stubble flecked his jaw. His hair was lank, flopping over his forehead. Pale, spidery lines were visible from beneath his shirt collar. I wondered if they were the scars from his throat cutting. He was wearing a blue striped tie, one he used to wear for work.

He looked up, and locked eyes with me. His lips lifted in a slight mocking sneer. I averted my gaze, remembering Peter's words about not looking at my father.

I thought my heart was going to explode out of my chest.

A stream of people were taking seats on another elevated timber platform to the side of the front bench. There were men and women, some old, some younger; they all looked ordinary. I assumed this was the jury.

"All rise," a male voice boomed out of nowhere.

A man in a robe and wig appeared from behind the panels in the grand wooden wall, absurdly reminding me of hidden rooms in fairy tales. His wig seemed to be fuller and longer than the others. A purple sash was draped around his neck. He looked regal. He sank into the single chair in the middle of the big bench. He must be the judge, I thought.

"Please be seated," the voice from nowhere boomed.

Everyone sat, hushed. A man in a black robe but no wig walked over to me. He instructed me to place my hand on the Bible. He asked if I promised to tell the whole truth and swear it in the name of God. Although not religious, I said I would. The words felt heavy with meaning as I uttered them.

Peter Brebner walked over to me and flashed a wide smile. He leaned against the timber panel and asked me gently to explain who I was and relay what had happened at the flat.

My heartbeat thumped in my ears. I felt slightly feverish. I leaned towards the microphone. I stumbled, fumbling with the words. I didn't seem capable of forming them. There was no spit in my mouth; my tongue felt double the size. I looked at him, feeling mortified. He smiled again, sympathy in his eyes.

"Take a deep breath, Melody. Take your time. We're in no rush here. Take a sip of water. When you're ready, start again."

I paused, slurped a mouthful of water from a glass, and took a deep breath. I started talking. This time, I felt my memories cascading. I closed my eyes and visualised myself back at the flat. I explained what I saw, what I said, and what I did.

He was watching me and, when I finished, he nodded. He then asked me questions about my father's behaviour at Brighton. He asked how it made me feel, whether I felt safe.

I concentrated on staying in control, pushing down simmering anger, sadness and fear. I wouldn't give my father the satisfaction of seeing me fail, I told myself.

My responses were abruptly cut short by a familiar voice from across the room.

"You lying little bitch."

I looked over at my father's crimson face. He was leaning forwards, hands on the desk in front of him, his eyes shooting daggers of spite in my direction.

"You're lying to protect your mother!" he roared. "You're a worthless, deceitful little bitch. You're always covering up for that woman you call a mother."

A loud bang came from the front of the room. I turned towards the judge.

"Mr. Horrill. If you don't refrain, you will be removed from the court. This is my only warning." He glared at my father. "Apologies, Ms. Horrill. Please, continue."

My father sank back into his seat, his eyes ablaze.

I was rattled. My heart was hammering so fast, I wondered if everyone could hear it. Tears threatened. I took a tissue from my pocket and dabbed at the corners of my eyes. I took another deep breath. I felt the air quiver as I exhaled.

Peter Brebner prompted me to pick up where I had left off. I talked about the day we were leaving.

"Liar!" my father shouted.

Another bang.

"Remove him!"

The two uniformed men escorted my father out of a side door. He glanced back over his shoulder at me. I didn't look away.

"Ms. Horrill," the judge said.

I turned. "Yes, Your Honour."

"Given your father's outbursts and abuse, you don't have to continue your testimony here today."

I paused. I looked at Mum. She was wiping her eyes with a hanky.

"No, Your Honour, it's okay, I'm okay. I'd rather get it over with."

"Very well then," he replied.

Peter Brebner finished questioning me. With my father out of the room, I felt better, less anxious.

"Your witness." He turned to another man seated at the front table who rose and walked over. My father's lawyer. He stopped in front of me and smiled. He was a big man, tall, with grey-blue stormy eyes. I summoned my defences, attempted to fortify my inner wall.

He politely asked me more questions about what had happened at the flat. He wanted specific details. He asked if I saw anything at my father's hand when he attacked my mother.

"Your father asserts that he didn't realise he had a corkscrew in his hand. He admits to slapping your mother but he says he didn't grind the corkscrew into her face. That was an accident."

In my mind's eye, I again saw my father's forearm bulging as he held my mother. I saw his jaw clench with exertion as his hand turned. I described what I remembered. The lawyer studied me, his face expressionless.

"So, Ms. Horrill. How many turns did it take to unscrew it?"

I was shocked. I looked at the floor, searching my memory. I wasn't sure how to respond. I had no idea. I felt panic rise, but this quickly gave way to indignation; to me, it was a ridiculous question.

The room stilled. I felt everyone's eyes on me. I looked up and stared at him.

"I don't know. I wasn't counting."

He turned away. "That's all, Your Honour. The defence releases the witness."

"You may leave, Ms. Horrill. Thank you," the judge said.

I rose from my chair. The clerk came over. As I walked out, I searched for the comforting face of Peter Dunstone. He was sitting in the front row. He smiled and nodded slightly. The clerk instructed me to bow as we walked through the door.

Back in the meeting room, I slumped against the hard back of the couch, feeling drained, empty and detached. My mother walked in and sat beside me. She reached for my hand. I couldn't find anything to say to her. I wanted silence. I wanted to go home. I wanted a drink and a cigarette.

A short time later, Peter Dunstone walked in.

"Time for lunch," he said, smiling, a warm light in his eyes. "After lunch, there will be closing arguments. The jury will then take time out to deliberate."

His tone seemed lighter. "Mel, I'm so sorry about your father. Why don't I take you home?"

I nodded.

"Doreen, you're welcome to wait in case there's a verdict. But, in reality, we probably won't get one today."

"That's alright, Peter. I think I'll just wait here for Sid. You go, I'll be fine." She sounded weary. I leaned over, kissed her on the cheek and walked out to the elevator with Peter.

He turned and hugged me. "I'm sorry your dad was so nasty in there, Mel. You did so well. I'm proud of you. It'll be alright, you'll see."

I broke the embrace and searched his eyes. I really wanted to believe him.

12

DROWNING
SORROWS

I CAN'T WAIT to get back on the river when I return to Adelaide after my brief trip to Queensland. I desperately want to see Jock.

I feel the same familiar thrill when we turn the corner into his territory and he races up to the boat.

I ask Mike if we could go to my favourite mangrove channel so we could quietly slip in for a swim with him. He agrees.

We turn into the channel, cut the motor and drift slowly. Jock swims to the ladder and waits impatiently. I've bought a new mask and snorkel and am keen to try them out. I sit on the back of the boat, feet dangling in the water, next to the ladder as I adjust the straps of my goggles.

I keep an eye on Jock as I fumble and see him turn his attention to my feet. I watch as he opens his mouth and closes it over my toes.

I don't move. I hold my breath. I know his teeth are sharp enough to crunch through crustaceans and he could inflict a nasty bite if he chose to. But I feel no fear. I trust him.

The caress is gentle, as if he is feeling the contours of my toes with the tips of his teeth. I've played with kittens and puppies who'd gnawed at my fingers, I've even had a horse lick my palm, but this was completely different. It seems that Jock is using another, alternative sense of touch to check out what I am, what I am made of. It doesn't even cross my mind that he may playfully nip or chew. He is just so gentle—his teeth barely tickle my skin.

Once again, the wonderful bizarreness of this relationship dawns on me. I also realise how much I have grown to implicitly trust Jock and how rare that is, at least for me.

After several seconds he releases my toes and resumes waiting at the foot of the ladder. I lower my goggles, pop the snorkel in my mouth and slide into the water. Today it is time to play "hide and seek" again. The joy of the game and Jock's mischievous energy embraces and invigorates me. I forget everything else. I am completely present.

I feel rejuvenated after seeing Jock and am keen to visit the other dolphins after meeting shadows of them in Western Australia and Queensland.

As we drive out into the river, I reflect on Jock's behaviour and realise that he is probably doing something that all dolphins do—connecting through touch.

In fact, dolphins always seem to touch one another when they're together in pods. It appears casual, incidental. The stroke of a pectoral fin along a companion, the flick of a fluke against another. Sometimes, though, male dolphins can become rough and boisterous with females, especially when they want to mate. Whether it's gentle or harsh, they interact physically much of the time. This is why, in comparison, Jock's life seems so empty, so lonely; he seems like such an outcast compared to these dolphins who appear to need physical interaction with one another.

As I watch the pods leap and play in the pressure waves of boats, I feel alive, somehow restored and regenerated, awake from the emotional coma that has consumed much of my early life.

Mike, too, is delighted every time we come across a pod of dolphins, especially when we hook up with regulars such as Two Notch and Hook. He's known this pair for years, much longer than I've been involved in Mike's research.

Two Notch is a big chunky dolphin. His slightly crooked dolphin smile makes me think that, if he were human, he'd be one of those people whose grin immediately makes other people smile. Mike named him Two Notch because he has two beer-bottle-cap-sized chunks missing from the trailing edge of his dorsal fin. Hook is his friend, another big dolphin who is thundercloud grey, a little darker than Two Notch. They spend a lot of time hanging out together.

We watch them swimming from one place to the next, fishing and generally being silly. They favour a wide part of the river, and boldly approach female dolphins. They are the scallywags of the river, a couple of cruising Casanovas.

One day, we see them join up with a small pod of females. They start frolicking. Then Two Notch flips onto his side and joins one of the females. Bellies touching, they swim together fluidly in perfect sequence. Their flippers touching one another in what looks like an embrace. Their tail movements become synchronised. They slide together effortlessly, fluently, quickly. To me, there is a tenderness in their meeting. It looks beautifully serene. After several seconds, the pair break off.

Now Hook seems to be nipping another dolphin's dorsal. I wonder if he is giving her a love bite. Then comes the same intimate behaviour I'd just watched Two Notch engage in.

Mike lowers his camera. "They were mating."

"I thought so. I wasn't sure. So, that's how they mate all the time?"

"Yep. They have sex belly-to-belly, or face-to-face, if you like. It may help to establish relationships."

"But dolphins aren't monogamous, are they? So why does it matter?"

He grins. "Maybe they do it because it's fun. They do a lot of stuff just for the hell of it. They're one of the only creatures on Earth that spend so much time just playing and mating."

I learned something that day from watching their joyful freedom, lack of inhibitions and natural instinct to frolic. While they might be considered promiscuous in human terms, dolphins are one of the few animals that mate for more than the sole purpose of reproduction, even if the encounter is fleeting and at times forceful.

Aside from the rough behaviour, which I had never personally witnessed, I wondered whether dolphins could remind us to take time out occasionally to play, have fun, perhaps be a little mischievous and just live for the moment.

I thought back to a time in my life when I took the idea of having a good time and living it up to the extreme. My single focus was escaping reality but, in the end, it couldn't last. Running away turned out to be an illusion.

MARK BUZZED AROUND the kitchen, rustling up something to eat. I sat at the small wooden table and watched him, sipping a glass of wine.

A wave of affection washed over me. Five years older, in his early twenties, Mark was already a giant of a man, more than six feet tall, and solid but not overweight. On the outside, he seemed big and strong, but there was still a little boy inside who just wanted to be loved and accepted. Usually, he had the brightest smile, which was contagious.

Since he'd arrived home today, however, he hadn't been smiling. He looked pale, his eyes sunken, eyelids fleshy.

As soon as he walked in, I told him about the day in court.

"Bloody hell, Mel, that's heavy," he said. "So, when do you think we'll hear a verdict?"

"I don't know, maybe tomorrow. That's what Peter said."

Mark cranked open a can and plonked its contents into a saucepan. I got the feeling he really wanted to talk about something else. Did he want to change the subject? They were short-staffed at the hospital. He cared about his patients and often discussed them. He also frequently mentioned a fellow nurse he was keen on, grinning when he talked about her.

Now, there was no sign of a grin. "Want some soup and toast?" he asked.

As I studied him, it struck me that I hadn't bothered to ask him how he was feeling. I hadn't considered what impact the attack and trial were having on him.

"Are you doing alright, Mark?" I asked. "We never talk about what happened in England or at Brighton."

He stopped stirring and turned to me. His brown eyes glazed over.

"Mel, I don't want to revisit it. I see no point," he said sharply. "Our father could soon be put in jail for attacking our mother. That's not something I'm particularly proud of, so can we please just drop it?"

I chewed my lip in frustration. "Why doesn't anyone in this family want to talk about anything? Let's all just pretend none of it ever happened, and it'll be alright. Is that it?" I could hear my voice rising.

Mark put his hands on his hips. Soup was slopping and plopping in the pan behind him.

"It's just that there's no point, Mel," he said. "What's done is done. We have to carry on, put it to the back of our minds. That's all there is to it."

"That's fine," I said, feeling exhausted again. "Whatever you want to do. I'm going to bed. I don't want any dinner. I'm not hungry."

I slammed my bedroom door and sat on my bed, playing Australian Crawl on my ancient cassette deck. "Down-hearted" came on, my Christmas anthem. Tears rolled down my cheeks as I remembered the misery.

I felt alone. My mother was living with her boyfriend and hadn't checked in on me. My brother wanted to pretend none of this had ever happened. Kev, who was now my boyfriend, didn't understand because he had the perfect family. And I'd already dumped too much on poor Jan.

Could I, as Mark had suggested, ever forget or move on?

The following afternoon, the call came from Peter Dunstone.

"Listen, the verdict's been handed down."

I held my breath.

"The jury found your father not guilty of intending to cause grievous bodily harm, but guilty of unlawful wounding."

"What does that mean?" I asked. "Is he going to jail?"

"The judge will sentence him later, but, yes, he will be going back to prison. I think it's a pretty good outcome. Your testimony was crucial."

I felt lightheaded, giddy, a swirl of emotions. My father was going to pay for what he'd done. I wouldn't have to worry about him stalking us. Mum would be safe.

"Thanks, Peter," I said, feeling weirdly elated. "You've been so great. I'm going to miss having you around."

"Pleasure, Mel. You're such a strong young woman. Look, if you ever need me, even just to chat, just call."

THREE MONTHS LATER, in March, the sentence was delivered by Justice Graham Prior.

He said the jury had decided my father had not intended to seriously harm my mother, who had left him the previous year, but was satisfied he had wanted to wound her. Justice Prior told my father he was lucky my mother had not been more seriously injured by grinding the corkscrew into her cheek.

The judge said psychiatrists had found my father suffered from "morbid jealousy" after my mother had left him. However, he said my father had not yet fully accepted responsibility for the attack, and still blamed his wife. The judge said a psychiatrist found his personality obsessional and that despite the morbid jealousy, my father did not have a personality disorder nor need psychiatric treatment. The judge said he hoped my father would accept that his marriage was over and would respond well to supervision and counselling.

With time served already, much of it in the hospital after he had cut his throat, my father would spend about eighteen months behind bars.

AFTER THE SENTENCE was handed down, I rang Sid.

"Hi Sid, is Mum around? Have you guys heard about Dad?"

"Your mother says she has a headache, she's lying down." Sid's abrupt tone shocked me. "The police called. She went completely crazy, insane, yelling and screaming that your father should have been put away for longer. I've never seen anyone act like that. We had a hell of a fight."

I waited. "Look, Sid, Mum can lose it at times. She just flares up and goes off. Let her cool off for a while. She'll apologise and then it'll be like nothing happened."

"Well, I'm not going to tolerate that kind of behaviour," he continued. "She can't go off at me like that. I have a weak heart. She's irrational. I can't do anything about your bloody father. God, at one stage, I thought she was going to throw something at me."

"Sorry, I don't know what to say," I replied, as calmly as I could.

"Well, if she continues like this, she can move in with you and your brother, as far as I'm concerned," he said. "I know she's had a bad time of it, but I will not be disrespected in my own home."

"Okay Sid. Look, I've gotta go. Please tell her I called, and I'll try another time."

I hung up. Another mix of emotions. I felt sorry for Sid, who seemed like a decent man. Was my mother going to push his buttons, like she'd sometimes pushed my father's? I wanted things to work out between them. Most of all, I wanted her to be happy and secure.

The following day, I drove to their place after work. My mother answered the door. She smiled brightly.

"Hi, Mel," she said. "Come in. Cup of tea?"

"Yep, thanks," I smiled. Her mood seemed extra chirpy. I followed her in, took off my shoes and sank into the pillowy couch.

"Sid's gone to play golf," she called from the kitchen. "Everything okay?"

She emerged with two steaming cups.

"Yeah, I guess," I said, blowing on my tea. "Sid told me you two had a fight last night."

She eyed me over her cup. "Did he now?" she said. "Well, yes, we had a bit of a row. He apologised, so it's okay now. So, your father's staying in jail? Thank God for that."

"Eighteen months, apparently. That should make you feel a bit safer?"

"Well, I just hope he takes this time inside to think about what he's done, hey? You never know, he might come out a changed man."

I raised my eyebrows and said nothing. I sipped my tea.

"Oh, I've got something to show you!" She heaved herself up from the couch, walked to the dining table and picked up a piece of newspaper. She presented it to me like a trophy.

"We made the paper," she said.

I read the short story from the Adelaide newspaper about the trial and sentence. Just a few lines. The whole sorry saga stuffed into three paragraphs. I handed it back to her.

"You can keep it if you want," she said.

"Why the hell would I want to keep it? What do you want me to do, frame it?"

Her mouth fell open, but no words came out.

I put my unfinished tea on the side table.

"I just wanted to make sure you're okay," I said. "But you seem to be fine. I'm tired, I've had a long day and I'm going home. See you soon."

As I pulled into my driveway, a darkness came over me. I was stuck. Marooned by my own frustration, anger and bitterness. Something had to change.

I DECIDED TO TAKE my brother's advice and try to forget. I desperately needed to escape. I called Kev. I told him I was ready to take our relationship to the next level. I wanted to spend even more time with him and for the first time become intimate, physically.

Kev's world was exciting, a stark contrast to my own, and I catapulted myself headlong into it. The next few months became a blur of parties and boozing.

His family was wealthy and he seemed to have an endless stream of funds. They accepted me, and I enjoyed getting to know them. It was obvious they cared about one another deeply. I settled quite quickly into the role of Kev's glamorous girlfriend. Life morphed into one lavish dinner after another, followed by drinking with loud, back-slapping strangers until the early hours.

Kev was kind and thoughtful, and I cared about him. He splurged on me. I spent hours learning how to apply make-up properly to look the part. I watched as he gambled hundreds of dollars at the casino. I told myself I loved him and was happy. He told me he loved me and I believed it.

His parents opened another nightclub in the city. Dominating the room was a large polished wooden dancefloor. Rainbow lights and pulses danced across the surface. Rich velvet lounges hugged the space. It seemed opulent, excessive and magical.

Kev managed the club; he suggested I quit my job at the restaurant and work with him, so I did. I learned how to make cocktails, and spent my nights in a kind of trance, swaying to hypnotic beats, mixing exotic concoctions in silver shakers for impossibly glamorous creatures.

At the end of each night, we'd eat breakfast, drink Bloody Marys and throw back shooters at one of the restaurants or pizza bars nearby. I smoked until my chest hurt. We'd leave at first light, staggering onto the street like wounded soldiers. I'd stay at Kev's, sleep off the booze, get up in the afternoon and start all over again.

I only went home to gather clothes. I avoided my mother and brother, and only occasionally called Jan.

During one of our phone chats, Jan told me she'd met a guy who didn't approve of my lifestyle. He was religious, part of a strict church. She told me she couldn't come out with me anymore. I said I understood but, once again, I felt abandoned.

ONE AFTERNOON, several months later, when I did pop home, I found Mark sitting in the lounge. His eyes widened when he saw me.

"Mel, you look like shit," he said. "And you smell like a bloody brewery."

"Good to see you, too, Mark," I replied, deadpan.

"Sit down. You need to know that Blitz died," he said, matter-of-factly, without emotion.

I plonked onto the couch, staring at him. My darling blue budgie was dead.

"What? How?" I felt the sting of tears.

"I found him when I got home the other afternoon. I think the neighbour's cat managed to get into his cage. I buried him for you. I'm sorry. I tried to call you at Kev's."

A nail had been driven into my heart. Guilt washed over me. I should have been home. I should have made sure his cage was secure.

Mark's expression softened. "Mel, look. I know you've been drinking a lot. I know you're having fun but, honestly, look at yourself. You don't look well."

His eyes filled with tears.

"It's my fault," I said. "I should have been here to look after Blitz," I sobbed. "But, please, don't give me a hard time, Mark. You told me to try and forget. That's what I've been doing."

Mark leaned over and pulled me to him. He smelled familiar. I recognised Old Spice, the aftershave I'd bought him the previous Christmas. I felt the thick fuzz on his arms as they enveloped me.

I pressed my face into his shoulder and cried. He held me tightly, making soothing sounds as he rocked me gently back and forth. When I pulled away, I'd left black and red smudges on his T-shirt.

Mark was studying my face. "I've been to see Dad."

"What? What do you mean?"

"I've been to see him. In jail."

I couldn't respond. I sat up and squared my shoulders to him. "He says he's sorry, Mel. He seems to have changed. He is writing letters to Ana in England, from prison, and she's writing back. I think he's trying to reconnect."

I felt the dark fury rising. "Not possible! He'll never change. He's just trying to manipulate you. I don't understand, Mark! Why, why would you do that?"

"He's still our father, Mel. We can't change that." He spoke quietly, avoiding my eyes. "I want to forgive him. I need to forgive him. I think you should come for a visit. It might help you move on. I'm going again in a couple of weeks."

The thought of seeing my father was hideous, preposterous. Why would I go and visit the man who'd brought us so much pain? Why would I subject myself to that?

I looked at my brother. His moist brown eyes were imploring, pleading with me.

"Come on, Mel, please. Do it for me."

I rubbed my neck, my fingers moving rapidly in a vain attempt to knead unyielding flesh. Hell, I just felt so tired.

Eventually, I heard myself say, "Okay, Mark, I'll think about it. But if I do this, I'm only doing it for you."

I didn't believe my father had changed. As much as I couldn't bear the thought of seeing him, a small part of me wondered if Mark was right. Maybe I had to face him again. Maybe this would help me to move on.

Besides, I reasoned, he couldn't hurt me behind bars.

13

A MEETING
IN CAPTIVITY

EVERY TIME I INTERACT with Jock, I feel a little closer, more bonded to him. When I'm not in the water with him, I'm worried about his welfare. When I am lonely or afraid, I wonder if he is too. I wonder about what he does at night. I have read that dolphins sleep by shutting off one half of their brain at a time. The awake half allows them to consciously keep breathing. Where does he sleep? Is it in the mangrove channels where we play or does he stay near his boat?

Despite his separation from his own species, he seems unencumbered by what happened yesterday or what will happen tomorrow. There are no prerequisites or provisos. There seems to be no emotional baggage, no fear, no disappointments, no expectations or judgement. Jock doesn't care if I've had a bad day at university, or I've fought with my mother or that I'm still healing from trauma. To me, he seems to be just concerned about when we will get in the water, if I wear flippers, have the paddle or am amenable to a game of hide and seek.

As I grow to know Jock better, I realise he is very smart. He possesses great problem-solving abilities, which he displays in our games. He is cheeky, sometimes mischievous but always gentle. He seems to be acutely aware that Mike, the other research assistants and I are not dolphins, therefore not as well adapted to his world as he is.

I feel sad every time I watch Jock return to circling his boat, alone. But over time, this steels my determination to learn more about the dolphins, not just in the Port River but around the world and somehow help them and their environment. I know I have to make a difference one day. I also know that in some way nature itself is a healer. This revelation is also something I want to share.

One thing I am acutely aware of, though, is how the very thing I am growing to love—the natural world—can also be so terribly cruel. This particular day on the water, I find myself cursing the river and questioning my faith in nature.

A THICK WOOLLEN SCARF wraps tightly around me like a Zulu neck coil as we head out. I hope my new and bulbous beanie, made for the snow fields, will also help obstruct the tendrils of arctic breath grasping greedily at my flesh.

Days like this are rare. The mercury isn't forecast to exceed 10°c. On the water, it feels half that.

Today, we're looking for Billie. Over the past few years she's lost two calves. They hadn't survived weaning and we don't know why. We know she's given birth again recently, but we haven't seen her for days.

Everything is sepia. The water, the sky, even the mangroves seem muted and monochrome.

Finding any dolphins, let alone Billie, in this flinty light will be difficult.

With the wind and water whisking and whipping, nature seems intent on stalling our arrival. Eventually, we make it to Billie's favourite spot. It's calmer here, protected from the worst of the wind by the wharves lining the bank.

Peering through binoculars, we scan the area. Perhaps Billie will leap out of the water in front of us, launch into one of her astonishing tail-walks, and display her acrobatic prowess to her calf. In my heart, however, I fear we won't be treated to such a marvel.

Up ahead, we spot a solitary dorsal. It's unusual to see a fin without others. It has to be Billie.

We approach cautiously. We don't want to stress her and, if her calf is there, we need to keep a wide berth.

We inch closer. I hold my breath, praying to whatever deity will listen: "Please, let her baby be okay."

She is moving slowly. She isn't fishing or diving. She appears to be just mooching. She is alone. There's no second "puh," no small, dusty, dark grey form taking clumsy breaths next to her.

Mike and both know it, but don't want to verbalise it. Saying it makes it real. Billie's baby is missing. If a baby dolphin isn't with its mother, it's most likely dead.

We continue watching from the boat. The only chills I now try to ward off are those spreading inside me. I'm used to seeing dolphins exude energy and vibrancy. Billie seems flat and dim, her movements are sluggish, lethargic. She doesn't approach our boat for a bow ride, she slips past without acknowledgement.

"I think we should leave her be," Mike says solemnly. I nod and settle into the passenger seat.

IN TOTAL BILLIE gave birth to seven calves, of which two survived. Several died soon after birth. One became trapped in a water sump at a soda ash factory and had to be rescued, lifted *by a crane back into the river. One of Billie's two surviving calves copied her tail-walking.*

I thought a lot about Billie's resilience. Like Jock, she'd been entangled in fishing line many times over the years. How could she deal with so many of her babies dying, yet still bring new life into the world, and nurture and teach it, not knowing if it would survive? I doubted many humans could cope with that, myself included.

I hadn't always been courageous, I'd also been a coward. There was a point in my life when I'd chosen to escape and forget rather than face pain. When I did turn back and meet it head on, I was wobbly and unsure of myself; I certainly wasn't bold and fearless.

Some time after the birth of her last calf, Billie lost weight rapidly. Her condition grew so poor, rangers collected a blood sample from her. She had severe renal failure. The decision was made to euthanise her.

The world lost an amazing natural wonder. Billie had survived so much yet still brought unparalleled joy to hundreds of people, who lined the banks to watch her tail-walk. I felt a deep sense of loss but, at the same time, her struggles, hardships and resilience had taught me so much. They'd taught me to be brave.

DURING THE DRIVE to the prison a few weeks later, in the summer of 1988, my insides churned. As we arrived, they turned to spin cycle. I was beginning to wish I hadn't agreed to visit our father.

A tower loomed ahead of us. "What's that?" I asked Mark.

"That's a watchhouse. I think they have armed guards up there, keeping an eye on things," he replied, keeping his gaze on the road.

Waves of nausea surged and receded as we pulled into the visitors' parking lot. I stepped out and wrapped my arms around myself. Ahead lay a hodgepodge of structures, like someone had collected ruddy bricks, old stone and metal, and thrown them together across a sprawling paddock.

A few moments before, we'd driven past ordinary houses, shops, pretty parks and a lush green sports field. It was bizarre to see this place of captivity so close to normal life. I caught glimpses of antiquated facades among the cold, steel structures. Everything was surrounded by sharp-toothed coils of razor wire, which looked ready to lunge at and constrict any prey attempting to escape.

Mark grabbed my elbow and urged me forward.

"It'll be alright, Mel. Stop stressing."

We entered a large mission-brown brick building. A slightly acidic bleach smell hung in the air. Behind a glass partition, a uniformed man asked us our names. We handed over our IDs.

I was told to pass my handbag to another guard for scanning. He removed my lighter, cigarettes and compact mirror, and said I could collect them on the way out. We were politely ordered to stand still while another man quickly patted us down.

Yet another guard led us through nondescript cream corridors and past a row of cells containing men. Dozens of eyes assessed me. I kept my gaze on the floor. It reminded me of the dog pound Mum and I had once visited after Dougal died. All those eyes, watching us. But these were not friendly, pleading eyes. I fought an urge to run back to the car.

We arrived at a door.

"You've got half an hour," the guard said. "Your visit will be monitored. If you want to leave early, knock." He buzzed the door open. Mark and I stepped inside.

Four mint walls. A white laminated table and four plastic chairs. The room couldn't have been more insipid. The same bleach smell hung in the air. Mark and I sat next to each other, facing two empty chairs. My heart was racing. Mark grabbed my moist hand. Another buzz. My father shuffled into the room. His V-neck top and pants were the colour of hay. Pale raised streaks were visible across his neck. He looked like he'd gained weight. His gaunt face had filled out since the last time I'd seen him. His once-lustrous hair was cropped close to his scalp.

He was carrying a plastic shopping bag, which he placed on the table. He pulled out a chair and sat down. His eyes darted from my brother to me.

I stiffened, straightening my back.

"Hello, Mark. Hello, Melody. Good of you to come visit your old dad." His voice crackled, like it was emanating from an AM radio with poor reception.

"Hi, Dad," said Mark.

"Hello, Father," I managed.

He looked away, dropped his eyes to the plastic bag and pulled out a black book. He placed it gently on the table in front of him. Holy Bible. He caressed the cover with his palm.

Returning to the bag, he pulled out two more items. He placed them on the table next to the Bible. They were two wooden owls, identical, about 30 centimetres high. Each had large dark timber circles for eyes and a dainty carved diamond-shaped beak. A wooden wing swept down one side.

He placed his hand on one. "Moneyboxes. There's a slit in the top to put coins in. You can get the coins out by taking off the wing." He pulled it away, revealing a round hole underneath. "I made them for you two in woodworking class."

He turned to me. "You know, owls represent death and rebirth." His eyes looked strange. "I've been reborn, Melody."

His irises were twitching, moving slightly back and forth. His gaze couldn't seem to connect and settle into mine. It was disconcerting. I rubbed my temple. My head was starting to hurt.

He placed his hand on the Bible and rubbed his palm slowly up and down its black cover. "I've discovered Jesus," he said, talking to the book. "God and Jesus, our Saviour, has saved me. I've pledged myself to Him. He forgives all sins. I study the Bible here."

He looked up at us. His mouth curved but the smile didn't reach his eyes. He gazed back at the book like a lover.

"I go to Bible class. I'm learning so much. God is my world in here. I carry this everywhere. I will go to Heaven."

"I'm glad you've found some solace in religion, Dad," Mark said gently. "I'm sure it's a good thing for you. God is good after all."

Something inside of me snapped. My face started to burn.

I gripped the side of the table. The words came rushing out.

"Is God going to forgive you for hurting Mum? You almost blinded her, you know. How are you going to repent for that? How are you going to ask for forgiveness for everything you've done to Mark and me?"

"It's alright, Mel," he said calmly, almost soothingly. "I know you're angry, but He will forgive you, too. He will forgive you for siding with your mother all these years. And I will forgive her for breaking her vows to me and to God. You know we were married in front of God?"

The room felt stifling. I was trapped. I needed to leave.

"Mark, I've got to go," I said.

I stood, looking down at my father. "Dad, I don't know if you've really found God, or if this is just another one of your games. I sincerely hope you do ask forgiveness and, for your sake, I hope He grants it."

I strode to the door and banged on it.

"I'll see you outside, Mark. I need a cigarette."

The guard opened the door and I walked out. I didn't notice the eyes in the cells this time. I collected my mirror and cigarettes then signed out. I walked to the car and slumped against it. My hands shook as I lit a smoke and inhaled deeply.

My mind was whirring. What if my father had changed, and he had really found God? Was I the one who needed to forgive? How could I?

Then I started wondering, was I deranged like my father? Mark was a better person than I was. He was capable of forgiveness. He wanted to forgive and move on. He was getting his life together. Mine was out of control. I tried to push away the darkness lingering at the edge of my consciousness. Would I ever be free of these black thoughts? Or, as my father had so often told me, would I end up being nothing? A lonely, useless waste of oxygen?

I wanted to crawl into a deep hole and hibernate.

I took another drag.

THE FOLLOWING DAY, I set out for Aldinga Beach, about a forty-minute drive south of my house. It was a weekday, so I suspected it would be quiet. As I got out of the car, I could feel the sun's warm kiss on my face even though the morning was brisk.

Aldinga was one of my favourite beaches. Its blanched shore seemed to stretch forever. In the distance rose rounded hills planted with pines.

I picked my way down a narrow path between sage-green succulents, which clung to one another, trying to anchor themselves in the shifting sand.

Turquoise waves lapped at the shore. I took off my sneakers and walked into them, watching them cover my bare feet. I closed my eyes, sinking into warm sand while being caressed by the water.

Ankle deep, I listened to the rhythmic slapping of the waves. I inhaled deeply, tasting the brackish air filling my lungs. There was no one else around; I felt enveloped by a deep sense of peace and belonging.

After a few minutes, I walked back to the base of the dunes and sat. It was so good to see sunlight and feel alive again, not hungover. I realised I didn't want to continue my life of the night with all that living in darkness.

Then I thought about my family and felt lost once more. My mother had created a new life for herself and didn't seem much interested in mine. My brother had Lyn, his new girl-friend. He was serious about her. He would end up marrying her and pop out lots of kids. As for my father, if he had really found God, then maybe he'd changed, but I doubted it.

I lay on my back, feeling the sand's snug embrace. I couldn't visualise my future. I didn't want to marry Kev. I couldn't see myself standing at an altar, promising him I'd stay forever. I couldn't imagine myself having children. But I knew he wanted a family. I was being selfish by staying with him while having doubts about starting a family. What if I were a terrible mother? My father's constant reminders of how stupid and useless I was, that I would never make anything of my life, replayed in my mind.

Perhaps I could end it. That way, I'd never have to see or hear my father again. I could just disappear. The memories, anger, sadness would all dissipate into the universe. The struggle would be over. But I dismissed the idea, knowing I'd probably never have the guts to do it.

A loud screech jolted me out of my thoughts. I opened my eyes. Two seagulls were at my feet. I heaved myself up onto my elbows. The pair looked at me, squawking and squealing.

"I don't have anything for you," I said.

More screeching protests.

I watched them watching me. One of them was standing strangely, and I realised it had no feet. Its uneven knobbly red legs ended in nubs. It wobbled, having trouble balancing.

Normally, seagulls annoyed me somewhat, but my heart ached for this bird. While its friend had perfectly webbed feet, it had none. I was amazed it was still standing, despite having nothing solid to balance on. It hadn't lost its voice, its courage, its spunk. On stumps, the seagull filled my ears with shrill shrieks.

"Good on you," I said to the bird.

I stood and dusted off the sand. The gulls flew away, yelling at each other, probably complaining to one another about the poor choice of human to approach for food.

I walked back to the car, sat at the wheel, turned on the radio and watched the waves. My foundations needed rebuilding. I couldn't and wouldn't rely on my family, Kev or anyone else and had to stop dwelling on what I didn't have and focus on what I did. And that was a passion for English and a desire to make it a bigger part of my life.

Yet, I was afraid of going back to school—terrified of failure, of once again failing to prove my father wrong. But I knew education was the only way for me to start building my bedrock.

I couldn't get home quickly enough.

An hour later, I pulled into the driveway and noticed Mark's car. To my surprise, my mother's was there, too. Inside, they were sitting at the kitchen table with mugs of tea in front of them.

"Hi, Mel," Mark said warmly. "I was just telling Mum about our visit with Dad."

"Oh, yeah. I don't think I'll be going back in a hurry," I replied.

"Well, I'm not going to force you, Mel. By the way, I left your owl moneybox in your room. I didn't know if you wanted to keep it or not. It's really well made. He put a lot of work into it."

"I guess it's about the only decent thing he's ever given me, so maybe I will keep it."

I sat with them. My mother's eyes had dark circles under them. Her lips weren't tinted with pink gloss. I spotted a grey runway cutting through her dark hair.

"You okay, Mum?"

She sighed. "Yeah, Sid and I have been a fighting a bit, that's all. He's just so fussy. I'm finding it hard to live up to his high standards, I suppose."

Sarcasm nipped at my tongue. "Well, I guess you could always move out, you know, be independent. But then, you'd have to fend for yourself."

She shot me a warning glance.

No one spoke. The silence was charged. I looked at my brother. To my surprise, he locked eyes with me.

"I'm going to change the subject," he said brightly. "I'm going to move in with Lyn. I think I want to marry her." A wide grin spread across his face. I knew he'd been seeing Lyn but this was happening more quickly than I thought it would.

"Wow, that's great, Mark." I gave him a big toothy grin. "That's awesome. I can't wait for the wedding. Make sure you get some decent music for the reception. I know some good DJs."

"Typical Mel." He rolled his eyes. "Always thinking about the party. So, what are you going to do? I won't leave you in the lurch. Maybe you can come and live with Lyn and me."

"Thanks, but no bloody way," I chuckled and winked at him. "I don't want to cramp your style. Actually, today I decided I want to go back and do Grade 12. I don't know how, but I need to. I can move in with Kev for a while."

Mark's face lit up.

"That's great, Mel. I think the community college offers Grade 12 for mature students. Why don't you give them a call?"

"You don't think I'm being silly, going back?"

"No, it's the best thing ever," he replied. "You might even go to uni! Wouldn't that be great, the only Horrill kid to go to university?"

He pushed up the end of his nose. "University educated, darling," he drawled in a posh voice.

We all laughed.

"What do you think, Mum?"

"Whatever you want to do, Mel," she said. "I'm happy for

you. I'll support whatever decision you make." Her voice was hollow. The words sounded like lines mothers are supposed to recite.

A week later, I enrolled at the community college during the mid-year intake. I could do my Grade 12 subjects over one or two years. I thought I'd try for one year, knowing I could roll a couple into the following year if I needed to.

Kev agreed to me moving in. I was grateful to him. I told myself I wasn't making a long-term commitment; it wasn't forever. I could leave when I wanted to, no strings attached. I did, however, feel guilty for keeping these thoughts to myself. I told him I needed to cut my hours at the club to Saturday nights only, because I needed to study. He grumbled about finding new staff.

Whenever I thought about going back to school, the tangled ball of anxiety that had seemed to live permanently in my tummy unravelled a little, and became easier to manage. It was my ball, I told myself; I owned it, and I'd just have to deal with the feeling. Eventually, I hoped there'd be no ball at all.

I finally felt a little more optimistic. My father was in jail and I was free to start building a new foundation. I just had to trust that I had the tools and smarts to do it.

THE FIRST FEW MONTHS after resuming study were a struggle. By now nineteen, and the youngest in the class, I was reassured to see middle-aged people studying, so I stuck at it. The months passed, my grades improved and I began to really enjoy English and even biology.

One afternoon, when I returned from class, the phone rang at Kev's. It was Mum. She and Sid had been having more problems, and she wanted to move out. She had applied for low-income community housing. But the phone call was not about the flat.

"You should know, Mel," she said woodenly, "your father has been released."

My grip on the phone tightened. "What, when?"

"Yesterday. The police told me."

I went to the couch. I needed to sit. "Where is he?"

"He's back at the house. I think Mark's going to see him this week."

"Oh God." My voice cracked. "We can't go through it again." I suddenly felt very tired.

Mum continued, sounding unconcerned. "Your father may not be a problem. Apparently, he was diagnosed with paranoid schizophrenia in jail. He's on medication now. Your brother says it seems to be helping."

"Really?" I was incredulous. "So when he was sentenced, the judge said he wasn't insane just insanely jealous and possessive. But now he is schizophrenic? Has he had schizophrenia all these years and they didn't pick it up or did he develop it in jail?"

"Mel, I don't know much about the disease, except that people hear voices, and they can get angry and deluded. It's possible he may have had it his whole life."

"Right. That's that then, isn't it?" I snapped, regretting it immediately. "I'm sorry, Mum, but this is a lot to take in. I knew he was due for release soon, but I didn't know when. I didn't mean to growl at you. I just don't know what to think."

"Well, maybe you should talk to your brother. He's a nurse and knows a bit about schizophrenia. I just thought you should know, that's all."

That was it. We hung up. I put my head in my hands and ran them through my hair.

None of this made any sense. I couldn't believe my father was mentally ill. I'd always felt he was in control, that he knew what he was doing.

OVER THE FOLLOWING MONTHS, Mark gave me regular updates. He spent a lot of time at the house in Brighton with Dad. There was good medication to control schizophrenia, he said, adamant it was helping my father. My father was keeping busy, tinkering with projects in the shed. He'd also started an "anti-divorce" club, holding regular meetings with other divorced men. At these meetings, apparently, men vented about their wives, and discussed God and the importance of not breaking marriage vows. I was shocked, but Mark thought it was harmless and gave my father something to do.

He also told me my father was fighting my mother through the Family Court. I guessed it may have had something to do with the divorce settlement.

Then, for some reason, Mark and Kev both started hassling me to see my father again. They claimed it would be good for me to see him out of jail and on medication. They told me I needed to close things off with my father, so I could move on. Their constant badgering was exasperating.

Finally, I told Kev I'd go, but only if he drove me there and waited in the car. He wanted to come in with me, but I told him if I was going to do this, I needed to face my father alone. I asked Mark to organise a time.

14

ONE
TORMENT ENDS

IT'S BEEN ABOUT two and a half years since I started visiting the dolphins. Because I've come to love being on the water, I have got my boat licence, coastal navigation, competent crew and diving certificates.

The dolphins and Jock have become my family and I recognise many of them easily. Jock's dorsal fin is so distinctive, it's unmistakable. On many occasions, as he hangs beside me in the water, I examine his disfigured fin, making sure I am gentle as I run my hand down it, feeling the bumps and bulges. My fingers find the creases and crevices where fishing line has bitten into his flesh leaving ugly white marks. Jock never seems to mind me touching his scars.

One morning, floating in the warm water I trace my fingers down his dorsal. Where it connects to his body, my fingers enclose the front of it and Jock suddenly takes off and I hold on. I can feel the water being displaced under me by the force of his powerful tail moving up and down. After a short distance I let go. He immediately turns back to me, dives and quickly surfaces right beside me. He lies still, his dorsal sitting next to my hand. I take the hint and again gently wrap my fingers around the front of his fin. Once more, he speeds off. I hold on. I am being pulled through the water at such speed it is surging and splashing in my face. I find it hard to breathe. I break off. We repeat this a few more times before Jock tires of the game and we return to the boat. Once again, I am overwhelmed by his trust in me and his desire to invent another game. But I am also acutely aware that dolphins need their dorsal fins to help keep them stable in the water and his seems so precarious that I feel we shouldn't play that game again. We never do.

Jock isn't the only Port River dolphin with an imperfect dorsal. In fact, I have come to recognise many of the residents by the nicks and cuts on their dorsal fins. Mike names some of the dolphins after the injuries to their fins. (Later, he even names a dolphin after me.)

One of my favourite dolphins is an old female called Mangled. The back of her dorsal fin is tattered, resembling an old saw with jagged teeth, perhaps the result of boat strikes, fishing line entanglements and fights with sharks.

Another favourite is Crew Cut. She's earned her name because the top of her fin is completely missing. It looks like it has been sliced off by a sword but, in reality, it was most likely a blade from a boat motor.

Later, she becomes a mother. We see her regularly and watch her care for her calf, which, unlike its mum, has a perfectly smooth, unblemished little fin. Crew Cut doesn't socialise much with the other dolphins. She appears to be content just looking after her youngster. On some days, we watch her chase after silvery snook or mullet, her damaged fin slicing effortlessly through the water. Initially, her calf hangs back and watches. As the months wear on, the youngster starts joining her mum.

One breathless morning, the summer sun is working its wizardry. The water glistens amber, turning into a gilded satin sheet. Dolphins are easy to spot on days like this, especially those with unusual fins like Crew Cut. As usual, she is swimming with her baby in the upper reaches of the river. We cut the motor and watch the pair swimming calmly. Without notice, Crew Cut speeds off. Flickers of silver skim the surface ahead of her. Spooked snooks. Her baby takes off after her, its little tail moving up and down as fast as a piston to keep up.

Crew Cut seems aware of her youngster's efforts and slows down, allowing it to catch up. She whacks the water with her strong tail; her baby copies her. Its small flukes can't match the size of mum's ripples, but it doesn't seem to matter. The tail slapping and chase last just a few minutes. It ends with mum and baby arching their backs and disappearing into the depths, presumably to feast on the stunned fish. I feel happy. I've witnessed a mother teaching her baby how to fish, how to survive.

Being an industrial hub where anglers vie with dolphins for fish, the Port River can be a harsh, often dangerous, environment for dolphins. I wonder how mother dolphins teach their young about the realities of living in such a place. Crew Cut has been teaching her offspring basic coping skills. How to hunt, where and what to eat, where to go, even perhaps how to avoid boat motors and keep your fin intact. I feel sure there are many more lessons to come.

I witness many interactions between mother dolphins and their calves. Each time, I am in awe. They seem to be devoted parents, skilful at tutoring their calves. As I watch Crew Cut and her baby reemerge and swim peacefully away from the boat, I ponder parenthood and my own life again. I wonder if I could ever teach a child how to successfully navigate life. No matter how much I am healing from my past, learning from the dolphins and exploring my own journey, I realise I would always be afraid of repeating the past. As I had reflected on that day at Aldinga when I decided to finish my education, the risk of exposing a child to what I'd been through was too great. That was the moment I confirmed my decision not to have children.

I also came to realise that, no matter what I did, I doubted if I could have actually helped my father. Hell, I wasn't even sure I could ever really forgive him. I finally understood; I was never going to receive the kindness and acceptance I yearned for. Some wounds left scars that I, like the dolphins, would simply have to learn to live with.

THE MORNING OF the visit arrived. Kev waited in the car as I walked up the driveway.

I felt like a little girl again. I looked at the same green weatherboard house, the same entrance, the same steps to the front door. My mother's beloved flowers were now shrivelled clumps of dried twigs. The once-lush lawn was barren, pock-marked with patches of dirt. But I was relieved to see my gum tree hideaway was still there at the side, strong and glorious.

From the car, Kev gave me the thumbs-up. I walked up the steps and rang the doorbell.

My father opened the door. He was wearing the same beige bomber jacket he'd always worn, the same belt with the yellow measuring tape. I stared at the belt; it had inflicted so much pain on my brother and sisters.

He'd grown a moustache. Grey, wiry and unkempt, it fell over his top lip. I couldn't see any auburn in his hair. He wore glasses with tortoiseshell frames. His skin seemed to hang from his jaw. He looked ancient but, in fact, he was only sixty-four years old.

"Hi, Dad," I said flatly.

"Hello, Melody," he said. He reeked of cigarettes. "I'm just making a cup of tea. Do you want one?"

I followed him into the lounge. The strong smell of stale tobacco hit me.

Two pink kitchen chairs sat in the middle of the lounge, facing each other. They were the same chairs I'd sat on at the kitchen table, a lifetime ago it seemed. There was no other furniture. I wondered why, but thought I shouldn't ask. It didn't feel like a lounge, more like an interrogation room.

"That'd be nice," I said, nibbling my nails. I sat on one of the chairs and crossed my legs.

Suddenly, I didn't want to be there. A voice in my head started screaming at me to leave. I took a deep breath, telling myself I was being a drama queen.

My father returned with two cups of tea. He sat in the chair opposite me.

"So, Melody, how have you been?" he asked, blowing on his tea. His eyes weren't jittery anymore. They locked firmly onto mine.

"Fine, thanks. Good. Actually, I've gone back to school. I'm almost finished."

The polite exchange felt false and surreal, like a scene in a B-grade movie.

"Yes, your brother told me." He smiled. "Well, maybe there's hope for you yet."

I bit my lip, telling myself to stay in control. "Mark tells me you've started an anti-divorce club."

"Yes, I think the legislation needs to change, so women can't just walk out on their husbands. They can't just break their vows because they want to. Divorce should be illegal."

I laughed. "Well, that's ridiculous. People get divorced all the time. You can't take people's rights away because you want to."

He squinted at me. "Your mother broke her vows, which she made before God. She shouldn't be allowed to do that." A familiar rage flared in his eyes. "How is she, anyway? Still with that Sri Lankan accountant? Is she still living at Seaton?"

"I don't know," I lied. "Look, Dad, I didn't come here to talk about Mum."

I put my cup on the floor. "You've got to let it go, Dad. She's never coming back. You two were like oil and water. It's better this way. You need to accept it. Let her go."

I found myself back in that garden in Cornwall, when I chose to go to my mother instead of him. I saw the old look of disdain, the bitterness in his eyes.

I swallowed my breath as frustration rose. Then, abruptly, I stood.

"Dad, you don't scare me anymore. I came here to try and make peace with you. But you just won't let me, will you? You won't let go of the past. You won't say sorry. You're nothing but a hateful, twisted old man."

He rose and grabbed my arm.

"Let go of me!" I shook myself out of his grasp. "I should never have come. I don't want to see you again. Don't bother trying to find me. Don't contact me. And, for Christ's sake, leave Mum alone. Move on."

I walked out.

When I reached the end of the driveway, I looked back. He was standing in the doorway, staring after me. I reached the car and yelled at Kev to start it.

I sobbed, the pain and disappointment pouring out of me. I felt defeated.

I turned to Kev. "I hope you're happy now. Don't ever ask me to do that again. I never want to see him again."

FOR WEEKS, I struggled with guilt about the visit. If he really was mentally ill, why couldn't I have been more tolerant, more understanding? What was wrong with me? I felt empathy for a footless seagull, but couldn't for my own father.

Yes, I despised him. I was angry with him, but I also wanted to hear him say "sorry." I kept seeing him standing at the doorway, staring at me. A voice in my head told me I should have turned back. At the very least, I should have tried to understand him. I should have told him I needed to make sense of it all, and asked him to help me do that.

Not long after, when I was studying for my final exams I received the call. It was my mother. I expected it to be the obligatory check-in. It wasn't.

"Mel, you'd better sit down," she said.

She's going to tell me she's left Sid, I thought.

"Your father's dead." Her tone was flat. Her words sounded so matter-of-fact, I didn't believe them.

"You're not serious."

"Yes, I am. Your brother has just identified the body. He's being interviewed by the police."

"What do you mean? Was he in a car accident or something?"

"Mel, he committed suicide. He hanged himself from a tree, the one you used to play on as a kid."

My mind went blank. I couldn't think. I couldn't talk.

"Mel, are you there?"

"Did you go? With Mark?" My body started to shake. My insides were taut strings; someone was plucking at them, and they were quavering and reverberating through me.

"No, he went by himself. Your father left a letter. Mark has it. Perhaps it's best he tell you about that." A long pause.

"What, is there something else?" I asked, dreading the answer.

"Apparently, there was some stuff in the shed."

"What kind of stuff?" My throat tightened. "Tell me."

"There were apparently homemade weapons, all sorts of weird things. Some had our names on them. Your father had welded together bits of metal. I think he meant to use them on us."

As she spoke, I froze. I couldn't believe what I was hearing. I started laughing hysterically. But this wasn't funny, so what was wrong with me?

I hung up without saying goodbye. I curled into a ball on the couch. I was no longer laughing. Instead, I heard a loud wailing noise coming from within me. I couldn't get a grip. I wanted to lose myself somewhere, anywhere. I was unravelling.

I WOKE UP sometime afterwards. My throat hurt. I grabbed my car keys. Half an hour later I was at Mark's. His eyes were puffy. Dark stubble shaded his jaw. He led me to the lounge.

I told him I was so sorry, I should have been with him when he identified the body. He avoided my eyes, telling me he was a health professional and he'd seen death before.

Then he pulled a piece of tattered paper from his pocket. Wordlessly, he handed it to me. I recognised my father's neat, precise handwriting.

Mark and Melody, I want you to know how much I loathe you. You are not my children. You have never been anything to me. You are nothing, failures, disappointments. Your mother broke our home. She's bleeding me dry. She will go to hell. You will all go to hell. My only child is Ana. I have left her everything. I'm finished.

I looked into Mark's eyes and saw deep wells of pain. I pulled him to me and held him. He wept. His body heaved in my arms like a floundering ship. I was drowning with him, sinking into a murky abyss.

"It's over," I whispered in his ear.

15

FEAR OF
FAILURE

ADELAIDE'S PORT RIVER is a special place. Resident dolphins have lived there for years, the population hanging on despite the threats of pollution, entanglement and boat strikes. The challenge ahead is to ensure future generations survive and thrive.

For more than forty years, Australia—once a whaling nation—has been a world leader in protecting dolphins and whales. In 1980, the federal government laid down stiff penalties for killing whales and dolphins in our national waters.

Six years later, an international moratorium on hunting cetaceans was agreed, but this hasn't stopped several countries from continuing with the practice, originally under the guise of "scientific research." Dolphins have been killed, either ritually because they were plundering tuna, or for food. Iceland dropped its own "research" program in 2006 and returned to commercial whaling although this is now in doubt.

Today, dolphin drive-hunting—driving them together by boats into a bay or onto a beach—still takes place in Japan, the Solomon Islands, Peru and the Faroe Islands. Each year in the

Faroe Islands, thousands of pilot whales are killed for food, in a tradition dating back hundreds of years to when meat was scarce. Now, however, the practice means the species is becoming endangered. Old habits are so hard to change.

Each year tens of thousands of dolphins around the world become entangled in fishing nets and drown. Thousands more become entangled in litter and debris. In the past year, ORRCA, an Australian volunteer marine rescue organization, responded to almost two thousand calls to help mammals in distress. There's also the issue of overfishing and leaving little for other species.

There are many challenges. But change is happening, albeit slowly. For example, in 1990, dolphin drive-hunting ceased off the Penghu Islands in Taiwan. And as recently as 2008, the practice ceased in Hawaii. There is an increasing awareness about litter in the ocean and a push for biodegradable plastics which break down in the marine environment. Hearteningly, some countries, such as Mexico, are banning the use of marine mammals for entertainment.

Conservation groups are continuing to lobby, spurred on by the results of studies on cetacean intelligence.

For years now, experts in evolutionary ecology and psychiatry have wondered if dolphins and other Cetacea are at least as intelligent as humans, and whether—because intelligence has many dimensions—it's possible that humans are less intelligent in some respects than certain larger-brained mammals.

Another area of study is language. Human language consists of fairly simple sounds arranged in elaborate sequences, whereas the cetacean language consists of extremely complex sounds delivered in units. The equivalent of a whole paragraph of information may be conveyed in one elaborate sound. When I was out on the river, I often wished I had the means to analyse what a single raspberry or whistle from Jock really meant, or what he saw in his mind when he used his sonar to check me and other humans out.

The job of protecting these precious mammals is never-ending. When I first met Jock, little did I know how passionate I would become about protecting marine mammals in the years to come.

Before I could embark on my fight to help them, however, I had to learn more about facing my own perceived threats and battling my insecurities.

THAT JOURNEY TO comprehend and confront my deep-rooted anxieties started in the summer of 1990—the year I first met Jock. I was twenty-one years old and was about to start university. I had completed Grade 12 with good scores, in spite of my father taking his own life a few days before my exams. I decided on a degree in communications majoring in psychology. I had chosen psychology because I wanted to know more about what made us tick and hoped to gain a better understanding of myself and my family. I had no idea I would learn more, much more, from Jock and the dolphins than anything from a textbook.

It was late January and my hands were slippery on the wheel of my orange 1972 Datsun. The car seemed smaller than usual, cramped. I had butterflies and my heart was racing, pumping through my cotton shirt. It wasn't just the summer heat. As I drove the twenty minutes to campus, I was terrified. I wanted to turn back at each stop sign. I wanted to throw up. Even though my father was dead, I couldn't mute his words in my mind.

Instead, the voice in my head was insistent. I can't do this. I'm going to fail. I'm just not smart enough. People will laugh at me. Why the hell did I think I was good enough? What was I thinking? Oh God.

A few months prior I had moved out of Kev's place and into a small, rented one-bedroom flat. We were still involved, but our relationship had deteriorated after I announced I had been accepted into university. Kev argued it would change me and corrupt my mind. I disagreed.

My brother was now married. My mother was still with Sid, but that relationship was also rocky. She had secured a community housing flat and would be moving in soon.

I pulled into the student parking lot and sat, eyes closed, trying to control my breathing. After a few minutes, to my relief, my heart slowed down. I grabbed my textbooks and set off for the psychology department.

After some navigating, I found the psychology building, climbed the stairs and entered a very 1970s-looking corridor, with a dull, grey-tiled linoleum floor and cream paintwork. The air smelled weirdly familiar. It had the same musty, old-book, white-sandwich, over-ripe-apple quality that I remembered from school. The scent made me queasy.

I was early. A handful of other students were wandering around, checking room numbers and glancing nervously at their watches. They were all a few years younger than I was and well dressed, sporting the latest designer jeans and shorts and trendy tops. I felt inadequate in my thrifted denim and eighties-style shirt.

I needed to sit down. I sat outside an empty classroom in the brace position with my head between my knees, waiting for my stomach and heart to calm down. I fought the urge to go home and forget the whole thing. I needed to do this, I told myself. I had to be strong. My father was wrong; I'd prove it.

As I bent over, mumbling to myself, I was sure people were staring at me. I didn't care, I needed to get regain control. Right, let's go, Mel. You can do this.

I stood and went to my first lecture.

A dozen students were already seated. I found a desk near the front, the uncool part of the classroom where I'd always felt comfortable. The cream Formica desks fitted the seventies vibe of the building. The walls were bare, apart from a projector screen and white plastic wall clock. In a corner stood a large whiteboard covered with illegible scribble.

I looked at my watch. This lecture should have started fifteen minutes ago. No one spoke, but everyone kept checking the time. The only sound was the intermittent rattling of the air-conditioning vent above the whiteboard. I found this strangely comforting, something to concentrate on in a weirdly quiet room.

The door flew open, breaking the silence. It rebounded off the rubber wall plug and back onto the man trying to enter, knocking him as he tussled with his stack of books. He stumbled into the room, looking hot and stressed. His round face was flushed, his glasses askew on the tip of his nose. His greying, wavy hair was in disarray. His short-sleeved blue shirt was dark at the armpits. He looked extremely uncomfortable. I liked him immediately.

He dumped his pile of books on the desk, smoothed back his damp hair, pushed his glasses back into place and surveyed the room.

"Sorry I'm late."

The man wiped his hands on his shirt. He was average height, stocky, with a wide barrel chest. His cargo pants seemed too big for him. He yanked them up over his hips. I watched beads of sweat snaking down the side of his face. He pulled a tissue from his pocket and wiped it away.

"I'm Dr. Mike Bossley, but I'd prefer you to call me Mike."

He paused, looking at the pile of books, then back at the class. The air-conditioning clattered on.

"Sorry everyone. Just getting my thoughts together. So, this is Intro Psych 1. I hope you're in the right class. In this unit, we'll cover the basics of the science of psychology and its principles. I'm also a marine biologist, so we'll be exploring the concept of humanity's connection to nature, or lack thereof."

Another pause. The vent rattled.

"I'm interested in this concept and I hope you'll find it stimulating. As you can probably tell, I'm not completely comfortable with lecturing, and I don't think I'm actually that good at it, but I'll do my best to encourage you to think and learn. You'll have a paper to write at the end of this unit; you must get at least 55 per cent to move on to Intro Psych 2. Any questions, ask me later. Now, let's get on with it."

I smiled to myself. This wasn't the kind of lecturer I had been expecting. Strangely, his discomfort comforted me, made me feel more at ease.

As he explained the essentials and principles of modern psychology and psychoanalysis—Freud's subconscious drivers, Maslow's hierarchy of needs, Jung's butterfly effect—his delivery was frequently interrupted by throat clearing and long pauses. He stopped frequently to mop his face, push his glasses back into place and locate information in his stash of books.

I noticed his bright blue eyes. While the rest of him seemed ill at ease, his eyes sparkled.

When the session finished and the class left, I tentatively approached him. He was trying to rebuild his stack of books. His face was flushed with frustration.

"Hi Mike, I'm Melody Horrill. I'm just wondering when you'll start talking about the nature connection. I think that'll be interesting."

He looked up from his pile of books, pushed his glasses back into place and said, "Ms. Horrill. What did you think of today's lecture?"

His directness caught me off guard.

"Um, it was a lot to take in but, yeah, kinda interesting."

"Nah, it was pretty crappy." A low chuckle escaped. "I always get nervous with new students and I'm not into the traditional psychology stuff as much as I used to be. I'm more into dolphins now. I think you'll like my next lecture more."

"No, I thought this lecture was good." I then rushed to make up for my apparent lack of enthusiasm. "Dolphins? I like dolphins. My mum gave me a book when I was little called *The Friendly Dolphins*. It was about how they saved people at sea and how smart they are. I've also seen schools of dolphins from a ship. So, you'll be talking about the nature connection next week?"

Mike looked at me. He took a breath and paused. "That's the plan. They're called pods, by the way, not schools. Dolphins aren't fish, they're mammals and live together in pods. I'll see you next week, Ms. Horrill."

Blood rushed to my face. How was I supposed to know what a group of dolphins was called?

With his books precariously under his arm, he shot me a quick smile and left the room.

I stood there among the empty desks, rattling vent and smell of stale sandwiches, reflecting on this awkward conversation.

THE NAUSEA AND HAMMERING HEART I felt while driving into university slowly subsided. As the weeks passed, I juggled attending daytime classes with a part-time evening course in public relations. I studied communication theory, writing and reasoning, statistics, and a few other core subjects—but Intro Psych 1 was what I looked forward to most.

To support myself, I'd got a job at a popular Italian restaurant in town, where I waitressed and tended bar. Sometimes, I managed to pick up shifts at a nearby nightclub. The hours were long and the work hard, but I liked interacting with people. The owners of the businesses treated me with kindness and respect. I'd also met a fellow student called Angelique. She also waitressed at the Italian restaurant, and we immediately hit it off. We discovered we were studying the same course at university. We grew to become close friends.

My focus, however, was squarely on succeeding at university.

Mike's second entrance wasn't nearly as awkward as his first. It might have had something to do with the cool change, which had descended on the city after a six-day-long heatwave. He wasn't anywhere near as frazzled as he had been the first week. He strode confidently into the room with his pile of books, and even managed a wide smile as he greeted the class.

He was right about me liking his second lecture: it was a remarkable mix of psychology and marine biology. He focused on humans' lack of connection to nature. He questioned our self-proclaimed position as the planet's most intelligent species and how this mindset has been used as an excuse to exploit it. He delved into our need to conquer and control nature, while forgetting that we are part of it. He explained how our dissociation with the natural world was messing with our heads and the planet, but asserted that we had the power to bring both back into balance.

His gestures were animated, his voice was richer, clearer and somehow louder. I enjoyed his passion and connected with the subject. I wanted to learn more. I felt comfortable, curious and invigorated.

During the second half of the lecture, he introduced the class to dolphins. He explained that, in many cases, our only exposure to dolphins was through visiting a theme park. Keeping marine mammals in concrete pools was inhumane, he said, but few people understood this. To us, they might appear happy, he went on, but research suggests they're not. The high-energy, joyful interactions dolphins have with trainers are simply tricks learned in exchange for food, the same as circus animals being forced to perform for crowds.

Mike explained that dolphins are acoustic creatures— they rely on sonar for navigation. But sonar bounces off hard surfaces like concrete. For a dolphin, living in a pool is like being forced to listen to yourself scream twenty-four hours a day, seven days a week. Eventually, it will drive you mad. It can make dolphins crazy, too.

Using the overhead projector, Mike showed us a series of images taken behind the scenes at overseas theme parks. Most distressing was a photo of an adult dolphin in a tiny pool, hardly wider than the length of its body. A blow-up alligator was tied to its dorsal fin to "keep it company." According to Mike, this dolphin had suffered severe psychological distress from being kept in captivity and become catatonic. Its sad fate was to spend the rest of its life motionless and friendless, bar contact with a plastic toy. That image affected me deeply; it still does.

Mike then changed tone, perhaps because several class members, myself included, were beginning to tear up.

He introduced Adelaide's Port River dolphins. He had been studying them for a few years. His interest had been triggered by a newspaper story about a dolphin swimming in the river alongside a racehorse and a dog.

Mike said he'd asked around and discovered that the horse belonged to Sandy, a trainer who regularly exercised his horses behind a small dinghy. His Jack Russell often joined them for a swim. Eventually, the dolphin—it was Billie—decided to accompany them. Mike met Sandy and his dog, and learned more about Billie who tagged along for the training session. This prompted Mike to invest in a small boat.

He started visiting the dolphins every week and studying them. Over time, he began to discern one dolphin from another by identifying the nicks and cuts in their dorsal fins. He documented where they were in the river, the behaviour they were displaying—fishing, mating, tail slapping or just cruising—and gradually developed a detailed picture of their society. He named the dolphins based on their physical traits. Graze had a wound on the front part of her dorsal, for instance, Two Notch had two notches in his dorsal, and Scarlett had a large scar across the base of her fin.

Mike described feeling fascinated while observing the dolphins' sentient, highly intelligent behaviour, complex social groups, individual idiosyncrasies, harmonious connections with the environment, and innate ability "just to have fun." He also suggested our species could learn a thing or two from dolphins.

During these lectures, I became immersed in the dolphins' world. I could see them in my mind's eye, watching their elegant, coordinated hunting expeditions. I imagined hearing the first intake of air of newborn calves. The Formica desk and plywood chair became my imaginary vantage point from Mike's boat.

Following one of these Monday morning "deep dives," I decided to make my move. I wanted to experience this world first-hand, not just conjure it up in my mind.

Mike was packing up his stack of books when I approached him.

"That was a great lecture, Mike. I really enjoyed class this week."

Not taking his eyes away from his books, he said, "Thanks."

"I was just wondering, do you take students out to visit the dolphins in the Port River?"

He paused and looked at me over his glasses. "Well, it's funny you should ask that, Ms. Horrill. I need another research assistant. I've had a few volunteers over the years, and have a couple at the moment, but I could always use an extra hand, if you're interested."

"Yes, of course, I'm interested. What do I have to do exactly?"

"Well, you'll meet me at the boat ramp when weather permits, early. You have to help me launch the boat, be a second pair of eyes looking for dolphins, help me document what's happening and, occasionally, drive the boat, or at least keep it steady while I'm taking pictures. I can't pay you. I barely get enough money from the uni to cover fuel, but you'll get to know the dolphins pretty well."

His words lit a fuse, activating me.

"Wow, that sounds awesome. What days do you go? I don't have a boat licence, but I can get one." I tried to control my excitement.

"Depends on the conditions. If there's a strong sou-easterly blowing, it's pointless going out, too bloody sloppy. But I can give you some idea a few days before. You'll probably get your ass wet, so make sure you bring a towel and a dry pair of pants. Yeah, a licence would be handy, but no rush."

I felt euphoric. "Okay thanks, that's great."

His eyes sparkled with amusement. "Do you like port?"

"What, as in the drink?" I asked.

"What else would I be talking about? Yes, the bloody drink."

"Yeah, I guess so, but I prefer a nice glass of red—why?"

"I keep a bottle of port on the boat. Gets pretty bitter out on the water in winter, which I'm assuming you'll stick around for, so you might need a shot of it to thaw out."

I wasn't sure how to respond. I figured he was just testing me out to see whether I had a sense of humour. I wanted to appear cool and in control, even though the little kid inside me was giggling with delight.

I decided to have a crack at a one-liner. "Well, as long as it's a tawny, it'll be okay," I grinned. "Seems appropriate to have a port at the port."

He looked at me deadpan.

I hurried on. "Okay, so when do we go?"

"Thursday looks alright, light sou-westerlies. See you at the Garden Island boat ramp at seven. It's a bit tricky to find, but you'll be right—unless you want me to pick you up."

"No, it's okay. I'll find it. But do me a favour, please, can you call me by my first name? Ms. Horrill seems a bit formal."

Gazing down at his books, he paused, studying a title on top of the pile. He looked up, the side of his mouth twitching like he was trying to suppress a grin.

"Right then, I'll call you Melo. Though I somehow doubt it's representative of your personality, it has a nice ring to it. See you Thursday."

He swept up his books and strode out, leaving me alone in the room, surrounded by vacant desks. Although this time, I felt sure I could make out a whiff of briny sea air.

16

MY LEARNING
BEGINS

WHEN THE ALARM went off at 5:30 am, I was already awake.

I wanted to be a research assistant but I didn't know if I was smart enough. What if Mike thought I was stupid or incompetent, and decided never to take me again? I had to impress him. I was so excited at the prospect of visiting the dolphins.

I downed a cup of tea and snatched my car keys. It was a forty-minute drive to the river.

The Garden Island ramp was a stone's throw from a landfill site, from which clouds of seagulls lifted in unison, hovering and squawking, while heavy machinery turned mountains of household waste into mounds. The odour of decay mingled with the spicy tang of salt seemed to cling to my skin, like the cloying dampness of the sultry air around me.

Mike was already there. He'd backed the trailer with the inflatable dinghy close to the water's edge and was fiddling with a small outboard motor.

"Hi, Melo. Just checking the motor. Can you help me push this off the trailer?"

I took off my shoes to help him launch the thick black rubber dinghy. It was surprisingly heavy. A small motor nestled between two rear pontoons. Two wooden slats provided extra seating, with two blue plastic paddles stowed beneath them on a plywood floor. A cooler-like box nestled in the front. I assumed it contained life-jackets and other emergency essentials, such as that bottle of port.

We hoisted the dinghy over the rear of the trailer and gave it a shove.

Although I'd seen the steam rising off the water, I couldn't believe how warm it felt.

"It's like standing in a bath," I said.

Mike was holding on to the opposite side of the dinghy, calf-deep in water.

"The water here's used for the Torrens Island Power Station. I'll take you to see it. It's just around the corner in Angas Inlet. River water is sucked in to run the steam turbines and the water coming out is about 24 degrees. Can you hear the hum?"

Just discernible was a low, mechanical, rhythmic wheezing sound. It was mildly disconcerting.

"Right, you hold on to the dingy. I'm just going to park the trailer."

"Okay, jump in," he said when he returned. "Let's go."

I hauled myself into the dingy and perched on the rubber side, as Mike yanked the outboard to life. The little motor sputtered. He twisted the arm, adding more juice. It began to rumble like a lawnmower.

"Gotta get this serviced," he mumbled.

It felt surreal to be out on the water with one of my uni lecturers. I'd never gone on trips with any of my teachers at school. I didn't know what the conventions were.

As we left the ramp, Mike turned left and headed towards the power station. We puttered along in silence. I hung on to the rope, surveying my strange surroundings. Dense forests of mangroves clung to the riverbanks. Their crooked trunks reminded me of wizened old men keeping watch over their territory. As the stench of the dump receded, the hum of the power station grew louder. In the middle of the channel, we passed a smattering of boats tethered to underwater moorings. Most had seen better days, being cobbled together with odd bits and pieces. Some resembled large floating blocks of plyboard with a small cabin perched on top like an afterthought. There were also newer small half-cabin boats, professionally made with their interiors open to the elements. All were covered with splats of bird droppings. They looked unused, unloved and abandoned.

"Does anyone use these boats?" I asked.

Mike chuckled. "Believe it or not, some people live on them." He pointed to one of the floating plywood blocks. "An old bloke and his dog live on that one. He built it himself."

Ahead, two towering chimneys appeared on the horizon, seemingly growing out of the mangroves. They must have been 150 metres high.

"The outlet for the sea water is just over there." Mike pointed to a concrete block embedded in the riverbank. No mangroves surrounded it. Instead of their intricate root systems, rocks had been placed either side to stabilise the soil. The outlet was spewing frothy water. Globs of white mousse floated on the surface.

"Does the hot water affect the dolphins?"

"A change in temperature can certainly impact the eco-system," Mike replied. "This is an industrialised urban area, so there's likely to be a lot of pollutants in the water."

He sounded so matter-of-fact, I was taken aback. I'd expected a bit more feeling. After all, the dolphins were his charges. But then, Mike was a scientist, so I figured he had to make sure emotion didn't cloud his judgement.

He turned the boat around, taking us back through the sad maze of forsaken moorings. I noticed a blobby clump kissing the surface a few feet away. It was moving around a boat.

A quiet thrill went through me. "That sort of looks like a dorsal fin. Is that a dolphin?"

My enthusiasm was quickly replaced by shock as I continued to stare at the grossly deformed dorsal fin gliding around the boat.

Mike slowed, then stopped the motor.

"Yep, that's Jock. He lives in this small part of the river. I've been watching him for a while. He's always here. For some reason, he likes to circle that particular boat. We know he's been tangled up a lot in fishing line and discarded nets and they've cut into his flesh. We think he's just a teenager in dolphin terms."

I felt so sorry for him; he looked so freakishly disfigured and had obviously suffered so much. He also seemed so lonely and isolated. I felt an immediate affinity, kinship and a deep empathy.

"Why don't you throw one of those oars into the water?" Mike suggested.

It was an odd request, I thought, but I picked up the paddle. It was light—only a few hundred grams. I hurled it away from the boat and it landed several metres away.

To my amazement, the clumpy dorsal broke off from circling and zoomed towards it.

Jock slithered under the aluminum shaft and manoeuvred his body until the oar lay perfectly balanced across the front of his dorsal. He swam off with the oar, controlling it precisely, like a tightrope walker with a pole. He wove around and through the boats, keeping his trinket in place all the time.

He disappeared, then returned moments later with the oar still in its spot. Eventually, he stopped and dived, leaving the plaything to drift. Then, with a flash of fluke, the paddle catapulted through the air. It hurtled in our direction, landing with a light smack a metre or so from the boat, where it bobbed on the water.

Jock returned to circling the boat.

I couldn't speak. I didn't know what to say. The sadness I'd felt moments ago when first seeing Jock was replaced with a bubble of chuckles percolating inside, bursting to escape. I couldn't remember the last time I'd seen anyone, or anything, have so much fun.

Then Mike said, "There's still quite a lot you'll learn about Jock—some of my other research assistants are already aware of him and know him well. I'm very particular about who I introduce him to. He's a special dolphin, and we need to try and keep him a secret."

I was confused but didn't let on. I guessed Mike would tell me in his own time.

So, this was the first time I had seen the dolphin with the deformed dorsal who would change me so profoundly. In our short meeting, I was already beginning to feel a powerful connection to Jock. Like me, he seemed to be an outcast, yet he also appeared to be seeking some kind of acceptance and attachment. I understood that too.

MIKE INSISTED THAT we try to find what he called "real dolphins" and leave Jock to his circling. I wanted to stay, but I knew Mike had serious research work to do and it was my job to help him. I didn't want to push my luck.

We picked up speed and left Garden Island in our wake. The river became wider, more expansive. The lingering dump smell gave way to the scent of the sea. The little boat bounced along. My bum was getting wet, but it didn't matter. With the wind whipping past me, I held the rope tightly, salty water flicking onto my face, moistening my lips. It was exhilarating.

Mike gave me a tour of the river. He pointed out the old boat-building yards, where only crumbling ramps remained. He took me past the soda ash company, which dominated one of the riverbanks. Covering more than a dozen hectares, it looked like a small city. Multiple sheds and red-brick buildings sprawled across the site. Chimneys protruded from the buildings, some billowing white smoke. Milky rivers, what I assumed to be factory waste, snaked into the river, blooming into chalky clouds as they met the water. I hated it. It all looked so ugly.

Going farther along the river, we arrived at a wide channel. To my amazement, it was littered with large shipwrecks, wedged into sandbanks between the mangroves. Their rusty hulls were splayed open, exposing huge decaying ribs. Unlike the power station and soda ash plant, nature had embraced these hulks. Cormorants and gulls perched on them. Mangroves sprouted from deep in their sinking bellies.

"What are these?"

"It's called the ship's graveyard. The government beached these ships here earlier this century. Apparently, it was cheaper than scuttling them. Some of these wrecks are almost a hundred years old. They're amazing, aren't they? The biggest one is the *Santiago*. I think it's from Scotland and hauled coal to the port back in the 1800s."

"They're beautiful. I can't believe they're here."

"Well, Port Adelaide itself used to be a busy working port. The locals called it Port Misery because of the mosquitos. Before that, the Indigenous people called it 'the Land of Sleep and Death.' I think it's amazing—it's one of a handful of places in the world where dolphins live so close to a capital city."

"I'm loving the history lesson, but are we going to see any dolphins?" I asked cheekily.

He smiled. "Let's go and find some."

Soon, we came across a pod. Mike slowed down, keeping a distance.

"Puh, puh, puh." Three dolphins emerged before taking another dive. After their descent, I noticed a perfect circle of still water on the surface.

"What's that?"

"Displacement. I call it a 'dolphin's footprint.' You can track where a dolphin is going by the footprints it leaves behind. Here, just hold on to the motor. Let's follow these guys for a while. Stay slow, keep your distance."

Opening the box at the front of the dinghy, Mike pulled out an impressive camera with an exceptionally long lens. He sat down, threw the lanyard around his neck, and started clicking. I tried to keep a slow, even speed, my hand gripping the outboard's handle tightly. He seemed completely absorbed.

"Did you see the notches on the dorsal of that big dolphin out front? I reckon that's Two Notch, which means Hook must be around here somewhere, too." He sounded excited.

Mike had identified more than two dozen dolphins by their dorsal fins. I squinted, trying to see the notches he was talking about.

"There's Hook!" he yelled. In the short time I'd known Mike, I hadn't seen him this enthused about anything.

"They're transiting," he said. "They don't seem to be fishing." He went back to the box, pulled out a notebook and pen, and started writing.

I watched the pod. They were so sleek, so agile. They glided through the water with no apparent effort. The dolphin called Two Notch pulled away from the pod, circled round and came over to the dingy. Immediately, I threw the motor into neutral—I was getting the hang of this little craft.

As he slid past us, Two Notch turned on his side. I looked down and saw his eye. I'd never seen a dolphin's eye before. I felt his gaze on me. I couldn't believe I was so close.

"G'day, Two Notch," Mike said, scribbling on his note pad. Two Notch returned to the pod.

"I haven't seen Two Notch in a couple of weeks. He's one of the first dolphins I saw when I started coming out here. I reckon he's an older male. He likes the ladies. I've seen him with loads of females."

I laughed. I hadn't felt this happy for a long time.

We saw eight dolphins that day. I was awe-struck with each new encounter, mesmerised by their beauty and grace.

Although I was an outsider, observing, I began to feel more at ease.

My first research trip ended in one of the narrow mangrove-lined backwaters, not far from the boat ramp from which we had launched.

Navigating the channel, Mike turned off the motor. We drifted slowly with the current. He pulled out a thermos and a pack of cookies. As we drifted with the low, wheezy hum of the power station in the distance, I felt like we were in the centre of an enchanted fairy-tale forest. I was gliding through a hidden corridor, part of a magical labyrinth hidden away from the rest of the world.

The Port River wasn't a mosquito-ridden, polluted swamp as I'd been led to believe. Sure, parts of it bore the scars of industry, but to me, it was divine, beautiful, lined with strangely exotic twisted trees where mystical creatures might lurk. Unlike nightmarish fairy tales, however, the residents of this world were magnificent beings: peaceful, ethereal and gentle.

One of them seemed extra special. I couldn't stop thinking about Jock. I felt drawn to him. Another wave of sadness washed over me. I wanted to communicate my compassion somehow, but I didn't know how.

I WOULD NEVER have believed that one day I would be jumping off another boat in this very channel and that Jock would not only play hide and seek with me, but that he would even come so close that I could stroke his back, tickle his tummy and that he would even nestle his snout in my hand.

17

JOCK FINDS
NEW FRIENDS

THAT MIKE HAD LET ME jump into the water with Jock within a few weeks told me that he trusted me to keep Jock's existence a secret.

So began a journey of trust that will forever be burned into my heart and my memory—and that would help me heal and reconnect to the world around me.

I can still remember the elation I experienced being in the water with this wild, beautiful yet damaged dolphin for the first time.

I can still feel his slick skin, in contrast to his rough snout, scarred mouth and knobbly dorsal.

To this day being received into Jock's world without conditions or enticement was, for me, extraordinary.

Feeling his sonar reverberate through my body as he examined and accepted me was an experience that was and still is humbling.

His unreserved acceptance and willingness to communicate both liberated me and held me captive.

Jock, the other dolphins and nature taught me so much and brought me a peace I could not once have imagined.

I had witnessed many wonders. I had watched in awe how dolphins and birds work together to fish, I'd seen births, deaths and a shocking case of human cruelty.

But soon I had to learn another tough lesson—sometimes you have to let go.

One day, after another successful dolphin-watching trip. I was helping Mike put the boat away.

He suddenly stopped mid-task.

"I've been thinking."

"Well, there's a first for the day," I joked as I closed the front hatch.

"I think we can try to lead Jock out to other dolphins. I don't think he's been socialised. He's an orphan, as you know, and probably was never taught. I think we can help him. He might make some new friends."

A warmth spread through me. "Yep, that's a great idea."

While I was enthusiastic about Mike's idea of helping Jock learn to socialise, at the time I had no concept of what that would mean for my friendship with this solitary creature. How could I know then that our plan to "enrich" Jock's life by helping him to meet new friends would cause me such personal grief?

NOW, ALMOST three years on from my first meeting with Jock and in my final months at university, I was doing well, getting distinctions in some of my subjects and would graduate. I got a part-time job as a receptionist at a car dealership and was still waitressing at the Italian restaurant on weekends.

I was in two minds about leaving uni. I had taken up Environmental Studies as an elective and loved it but wondered how to meld communications and the environment into a career.

I knew I would miss being Mike's research assistant and visiting the dolphins, particularly Jock. Mike assured me he would take me out when I wanted.

So now it came time to plan how we would try to join Jock up with dolphin mates.

One morning after our secret playtime, Jock followed Mike, Steve and I out to the wider channel. Although he had done this before, he had always turned back when we slowed down. We usually followed him, ensuring he got home safely before we left to visit other dolphins.

Mike opened the throttle and the boat rushed forward. As usual, Jock leaped in the wake, high in the air like an aquatic acrobat. Over and over he flung himself into the sky, belly flopping back into the frothy water. I felt the familiar thrill of watching him. This time, though, instead of slowing down, Mike kept going. We were now out of what we knew to be Jock's comfort zone and farther into the broader river.

We saw a pod of dolphins up ahead. Mike eased back and put the motor into neutral. Jock stopped leaping. Rather than turning around and heading home, he hung behind us as we drifted with the tide. The pod approached us. We recognised their fins. They were regulars in this part of the river.

I watched Jock. He was mooching behind us, like a shy teenager. The pod moved past the boat and approached him. He didn't move. I could hear his raspberries and sonar going off like a raucous car alarm. The group slowly circled him. They seemed to be checking him out. We watched. Eventually, we turned the boat around and Jock followed us back.

For several weeks we repeated this almost every time we visited Jock. We stayed longer each time. Jock gradually, tentatively, moved out of the boat's shadow.

One glorious spring morning the three of us launched the boat. The water was placid, a gossamer of steam lingering just above the surface. I'd seen this so many times before, but I was struck again by how ethereal, almost spiritual, it seemed. The low hum of the power station was the only reminder that we were a stone's throw from the industrial heart of a major port.

We drove to Jock's patch. He wasn't there. I felt anxious. Mike and Steve looked worried too. We searched every inch of Jock's territory. I was desperate to hear the familiar "puh" and to see that damaged dorsal fin tip slice through the water towards us.

Everything suddenly seemed unnaturally still. The tranquillity now felt tinged with trepidation.

"Where is he?" I asked desperately. "He's always here."

"I don't know," Mike said. "Maybe he's gone off to explore on his own. We'll come back later."

I didn't want to go anywhere. I wanted to stay and wait for Jock to return. Mike turned the boat around and headed down the wider channel. We'd been travelling for around ten minutes when Steve yelled to Mike to stop. There were dolphin footprints on the water, a few metres ahead. We waited for the pod to surface. Mike grabbed his camera. My pen was poised, ready to document the date, time, dolphin name and weather conditions.

"Puh."

I recognised a few of the fins and began to jot down details.

"Puh." I looked over and saw the disfigured dorsal.

"Jock!" I screamed. Mike shut off the motor. We allowed ourselves to be carried by the river.

Jock was touching the other dolphins. He was rubbing his body along theirs, nudging them as he'd done with me so many times. The others were reciprocating. I heard a cacophony of raspberries and sonars. We watched as he dived with the group and surfaced with them.

Then he swam over to our boat and barrel-rolled on his side. I stared into his unfathomable dark eyes. He circled once and broke off and went back to the pod. They left us, travelled upriver, and disappeared around a bend.

No one spoke. I looked at Mike and Steve. They were staring after the pod. I saw a tear form in the corner of Mike's eye. Steve blew his nose. My bottom lip was twitching.

"Well, I guess this is quite the day," Mike said, wiping his eyes. "Jock's found friends and he doesn't seem to want to know us anymore, so I guess we should go and find some other dolphins to watch. I think we might need a glass of port when we get back."

My heart ached. My best friend had left me. Mike threw the boat into gear, and we headed off. Later that day we swung by Jock's area again. He wasn't there. We headed in and sat on a patch of grass near the boat ramp, sipping port.

"Here's to Jock," I said, holding my plastic mug in the air. "He's finally been accepted by his own. I'm both happy and sad." We clicked cups. "Do you think he'll stay with them?"

Mike nodded, looking at the ground. "Probably. It's for the best. As much as I had doubts as to whether we should befriend him, I think we helped him reconnect with the other dolphins. We guided him out of his self-imposed cage."

I nodded. "I know, but I'm going to miss him so much. He's become such a big part of our lives. We can still check on him and his new pod, can't we?"

"Sure," Mike smiled. "You never know, he might find a nice lady dolphin and have babies."

That made me smile. "I'd love that."

DURING MY FINAL TRIPS out on the boat, Jock never returned to us again. As far as I knew he didn't return to his patch but I later learnt that Mike and the others continued to interact with him. When I did see him again, I watched him from afar, playing and swimming with his new friends. On one of my last trips out, he ignored us. Part of me felt heartbroken. I admit it—I felt abandoned again. I knew, though, it was a good thing for Jock. He might even have his own family one day. Maybe I would see a crumpled baby Jock with a dodgy dorsal pushed to the surface for its first breath.

I didn't venture back to the private playground. It seemed empty. The magic had gone. It felt like my dolphin trips became all business.

I finished my studies and graduated. I promised Mike I would stay in touch and still come out on the boat while I tried to find a "proper" full-time job.

Although I was happy that Jock had been accepted by others, I constantly worried about him. I fretted about what would happen if someone unfriendly approached him, tried to hurt him. The thought made me sick. I knew Mike had the same concerns. Other solitary dolphins around the world who had befriended people had been abused or killed.

Nevertheless, I reflected on how much I'd been transformed by Jock and the river. I knew he had helped to repair me. The experiences had let sunlight into my soul, and with it the knowledge that there was love and acceptance in the world.

I didn't want to leave this life behind, but I knew I had to move on.

Fate, however, had other plans.

18

INCONSOLABLE SORROW

MY WORLD SHRANK into a small, white-walled office, with a window looking out over one of the car lots. A few months after finishing university, a dealership where I'd been working casually offered me a job as the group PR officer. I accepted it. I bought an old, small cottage in an industrial part of Adelaide. It was my first real home.

I tried to enjoy the job, but I felt stifled and hemmed in. Often, I sat at my desk, flicking through paperwork, day-dreaming about being back on the water. Mike provided me with regular updates on the dolphins' activities and new calves he'd seen.

One afternoon in July 1993, I was packing up to go home when, my desk phone rang.

Mike's voice sounded snuffly.

"Hi, Mike. I'm just about to head home. I've got a bloody headache from all the customer satisfaction surveys. You sound like you've got a cold."

"I've got something to tell you. You're not going to like it."

A wave of dread surged in my stomach.

"What's wrong?" I whispered, as the office herd trampled past to the exit.

"It's Jock," Mike's voice broke. "He's dead."

I pressed a hand to my mouth, pushing back the scream that was about to break loose.

"Mike, where are you?"

"I'm at home."

"I'll be there as soon as I can."

I ran to my car and sped through the Adelaide Hills to Mike's house. My vision was blurred and I could barely navigate the twisting road. I pulled into Mike's driveway forty minutes later. We hugged, sobbing. I'd never seen Mike cry; his grief amplified my own misery. We walked into his lounge room. He poured two tumblers of scotch.

"How?" I asked, taking a swig of whisky to try to dull the pain.

"I don't know. The museum collected his body. He had a wound on his side and was quite decomposed. They're going to run toxicology tests. You know what it's like where he lived, it's pretty polluted."

"Oh God, my worst fear was he'd be hit by a boat or intentionally killed. And now…" I gripped the glass tightly and struggled on. "When will we know about the tests?"

"Maybe a week or so. I'm so sorry. I know how much he meant to you."

"He was my best friend." I tried a teary lopsided smile, but it slipped from my face. "He was irreplaceable."

I'd experienced death before—my father had committed suicide—but the depth of my grief for Jock was unlike anything I'd ever felt. This time the pain felt physical, like someone had ripped out my heart.

I don't remember the drive home. I took the rest of the week off work and spent most of it in bed. I rejected all offers of company. I needed to be alone.

Mike told me the results from the museum were inconclusive. However, later analysis of Jock's tissue found elevated levels of toxic chemicals called PCBs (polychlorinated biphenyls). Now banned, power plants and other industries used PCBs for heavy machinery. They remain in the marine environment for decades and can weaken the immune systems of mammals. They're also transferred from mothers to calves via milk. While the PCB levels found in Jock were elevated, experts couldn't determine whether they alone had caused his death.

While the actual cause of his death was a mystery, I couldn't shake the fear that we may have inadvertently put him in danger by helping him to trust humans. He had a wound in his side but that may have been from a boat propeller. But what if he had been attacked, poisoned, punched, stabbed or even shot at? What if someone poured something down his blowhole? These things happened, I knew. I felt sick at the thought.

WEEKS PASSED. I couldn't shake my melancholy. My work colleagues complained that I didn't seem engaged. They were right—I wasn't. Adding to my sorrow, Mike told me the university had cut funding for his research.

My despair morphed into anger, then resolve. One morning, I woke up with an idea. I'd be damned if Jock's death was going to be for nothing. We could start a charity to raise money for an awareness campaign about the dolphins and fund Mike's work.

The fortress inside me strengthened. I had to try, even if I failed. I resigned from my job—the company was probably relieved. Although I had a small mortgage, I had some savings and could go back to waitressing if I needed to.

I called Mike.

"We need to start a foundation. We'll start a dolphin sponsorship program. We'll create a newsletter, keep members up to date and maybe attract corporate sponsorship?"

"Wow, do you really think so? Can you afford to do this?"

"I can do the PR. We can enlist public support and educate people and kids. We can do school visits, you can do talks. I can probably get some news stories up. Besides, I've already quit so you can't say 'no.'"

Mike was on board. "If you bring in sponsorship money, you could take a small salary. We'd need a board and directors. I guess I could ask some of the other students like Steve."

"Yep, let's give it a shot."

For the first time in ages, I felt a surge of optimism.

WE FORMED THE Australian Dolphin Research Foundation, or ADRF. I became single-minded in my determination to raise awareness. Social media hadn't been invented yet, so I had to find ways of communicating to the public en masse.

I developed the "Sponsor a Port River Dolphin" program and organised a story about it with one of the local TV news services. We took the crew out on Mike's boat, and talked about how important the dolphins were, what threats they faced and how people could now sponsor one.

Angelique, my girlfriend from the Italian restaurant and fellow university student, wrote beautiful calligraphy. We spent hours sprawled on the frayed carpet in my tiny cottage, nibbling Yo-Yos and handwriting sponsorship certificates.

We developed three layers of sponsorship—standard, silver and gold. I went through the phone book and cold-called organisations, then followed up with a letter, asking them to become corporate sponsors. We attracted some big names.

Mike presented his research to community and industry groups. We spoke to people in pubs, playgrounds and parks.

I secured media sponsorship with a local TV station, which made free advertisements for us that they aired during downtimes. I fronted these ads, astonished to find myself before a camera voluntarily. The last time I'd interacted with TV news crews was back in the flat after the attack. It was uncomfortable, but I knew it would be worth it.

The station's much-loved veteran newsreader, Kevin Crease, and I visited schools in their news chopper. Sometimes Mike and Steve would join us. Kevin loved the ocean and sailing and quickly became a strong supporter of our work. Together, we spoke to hundreds of children about dolphins and the importance of looking after their world. One school near the Port River held regular fundraising drives for the foundation. One of the teachers, Elaine Everett, became a passionate voice in the local community for the dolphins and encouraged her students to learn more about the marine environment. Her energy, enthusiasm and support touched me deeply.

The Port River dolphins were beginning to generate a following. Journalists started questioning government leaders about pollution and conservation. Dolphins were becoming a ratings winner.

I secured a deal with a large bakery to promote the dolphins on its bread bags, with a percentage from every loaf sold going to the foundation. We organised regular paid trips on a graceful sail-training ketch called *Falie*. Built in 1919, it seemed as fused with the sea as the dolphins it enticed to frolic in its bow. We took dozens of paying passengers along the river and into the gulf, showed them the dolphins, and provided the opportunity to experience their world.

We also took out school groups. I marvelled at the children's smiles and squeals of delight when dolphins played in the bow wave. We were teaching a new generation how amazing these marine mammals were, and instilling in them an appreciation and love of the ocean.

I had found my voice and was using it, all thanks to Jock and the dolphins. People were listening. It felt as if the whole state was becoming aware of the special pods of resident dolphins right on their doorstep. While I could still sometimes hear my father's words in my head, I pushed them aside.

Was it really just a short while ago that I'd been sitting with my head between my knees outside the lecture room for the first time, trying to calm my stomach and slow my heart? I'd told myself then that I had to be strong and face my new challenges, that my father was wrong and I would prove it. Now, when I questioned my ability once again, I repeated to myself: Right, let's go, Mel. You can do this.

I focused on my passion, battling through my insecurities like speaking in front of crowds and addressing the media. I told myself I was brave and could do what I needed and wanted to do. I would take a leaf out of Jock's book and leave my comfort zone. With some training from Kevin Crease, I learned to look and sound confident—even if I was still trembling with fear inside.

With each new donation or sponsorship, my heart swelled

with pride. We recruited volunteers and ambassadors to spread the word and help with the growing workload.

A year or so later, to help pay the mortgage, I took a part-time job to promote the rollout of South Australia's first curbside recycling scheme. I loved the work. I co-wrote a children's book about how recycling works, which won a local award. Then the TV station that sponsored the foundation asked me to present segments on a daytime chat show about environmental issues and initiatives. Now, I was discussing everything from stormwater to worm farms. I became the spokesperson for the Catchment Management Board and fronted community service announcements about the dangers of litter ending up in the sea. At every chance I got, I promoted my first love—the Port River dolphins and the importance of protecting them and conserving their habitat.

With the dolphins' increasing profile and calls for the river to be cleaned up, I started receiving threatening phone calls from PR managers, working for industries that were known polluters of the river. While these calls rattled me, they only fuelled my determination.

The foundation attracted people from all walks of life who wanted to help. We organised a group of volunteers to help rescue a dolphin called "Float Baby," whose pectoral fins had become entangled in fishing line and net floats. With a united effort, they caught Float Baby and removed the debris. She lived with a large, unsightly tumour on her pectoral as a result. We renamed her Phoebe; she became one of the most popular dolphins on the sponsorship list.

Regularly, I went out with Mike on the water. Every time we passed the boat Jock used to circle, I thought of him and felt that familiar tug of loss. I'd remember our friendship and how much he'd taught me. I never shared just how much Jock had healed me with anyone, not even Mike. It was something I was unwilling to share; it was too precious, too intimate. The memories of our time together remained firmly in my heart, but I was slowly learning to live with the grief.

During these boat trips, I reacquainted myself with old friends such as Two Notch and Hook. I watched biker dolphins try to intimidate females in the river, and thought about how they'd inspired me to find my strength and stand up for what I believed in. I even saw Billie again, and wondered again how she had coped with five of her calves not surviving.

With fresh eyes, I appreciated the beauty and synchronicity of dolphins, as they cooperated with seabirds and one another. Teamwork would ensure the foundation's success, too.

Many times, I felt I'd bitten off more than I could chew, but then I'd think of Jock. The memory of his friendship and acceptance propelled me forward, just as his powerful tail flukes had propelled him and me through the water.

In the mid- to late 1990s, our determination increased when there was a spate of horrific attacks on the dolphins. Bodies started washing up on the banks of the Port River riddled with bullet holes. Who would kill a dolphin and why? Mike went into a frenzy of media interviews. I did some stories for the TV station and we pleaded with the public for help to find those responsible. It made national news.

Seeing my friends slaughtered triggered a deep rage within me. Every time I saw one of their glistening grey forms peppered with bloody punctures, I felt physically sick.

Many dolphins, young and old, were killed. The government increased patrols in the river. Eventually, the attacks stopped but the culprits were never found. The gruesome sight of my dead friends haunts me to this day.

AS THE CAMPAIGNING CONTINUED, I was asked to report for a TV wine show. It wasn't well paid, but I accepted. I needed to learn how to become a better communicator.

This led to a full-time career in TV news, reporting on environmental and science issues. I was also thrust into presenting the weather. The news directors told me it was a natural fit. Reluctantly, I agreed to give it a go.

Every contract I signed stipulated the station's ongoing sponsorship of the foundation, which was now largely run by Mike, the board and volunteers.

On TV I knew the part to play and the look I had to have, the voice I needed to use. I felt a bit like a ham actor on camera. Eventually, I realised I was much better off just being myself, even if I didn't fit the traditional TV presenter mould. People seemed to respond to it. I loved reporting. The stories were real and I revelled in telling them, especially when they involved helping the environment, wildlife and animals. Sometimes, I still struggled with anxiety, occasionally even on-air, and I went to counselling, which helped.

Then a news director at one of the stations, Grant Heading, asked me to come up with a concept for a documentary. I knew it had to be on Jock and the Port River dolphins. I asked Mike for the footage he'd taken during our times on the river.

Sitting in a small, dark editing room, I watched hours of tape, so much of it replaying my life with Jock. Feelings flooded back. I was that young woman at university again, yearning for acceptance and connection. I watched myself become immersed in my friendship with Jock and saw how happy I looked when I was with him in the water. I also noticed how eager he was to interact with me, Mike and the other research assistants.

I worked with a talented producer, Tony Morabito, and together we created an hour-long documentary called *A Dance With a Dolphin*. Months later, in mid-2001, we launched the film at a small cinema in Adelaide. I invited the Environment Minister and other officials. It received a standing ovation with few dry eyes at the end. I gave a short speech, thanking everyone for their support.

I felt so much gratitude, but also a sense of disbelief. This was unreal. From the stage, I saw Mike in the audience and knew he was proud of me.

The documentary aired across Australia to rave reviews. It made the front page of the local paper's TV magazine. In the US, CNN contacted me through my news director and invited me to its headquarters to present our work to the world. I was thrilled and spent six weeks in Atlanta, seeing the dolphins broadcast worldwide before I returned to Adelaide.

The state government contacted Mike, who, along with other passionate community advocates and dedicated marine campaigner Dr. Margi Prideaux, had been lobbying for better dolphin protection. The government agreed to designate parts of the Port River a dolphin sanctuary. It would be the first of its kind in Australia.

In 2002 the Environment Minister made the announcement on board *Falie*, the ketch on which we took visitors to see the dolphins. Three years later, the Port River Dolphin Sanctuary was officially declared. A global dolphin protection society eventually took over the work of the foundation. It continued to financially support Mike's research, and it took over the sponsorship program. The community groups evolved, keeping pressure on the government to continually improve conditions in the river.

During this time, I'd become a well-known presenter in Adelaide, reporting on the environment and science, along with regular dolphin updates. I won several prestigious awards for my stories and, one year, was even voted best presenter. But the truth is, I never felt completely comfortable on camera or hobnobbing with people who always seemed so confident. Part of me felt like a fraud, at times, still battling my father's legacy. Some high-profile women I worked with tried to bully and intimidate me. I'm told it was because they felt threatened. But channelling my precious female dolphins who faced off the bikers, I chose not to cower—I fought back. As a weather presenter, I learned more about nature and its remarkable power to both create and destroy. I attracted a loyal following and was humbled by it, but never felt I deserved the attention. Being in the spotlight was something I still struggled with; it was an internal conflict, one which I was determined to win. I felt grateful to the many viewers who seemed to like what I was doing and expressed their appreciation. I concentrated on animal and marine stories and kept up my regular Port River dolphin updates, which kept me grounded.

The dolphins were never far from my thoughts and, at every opportunity, I would do feature-length stories on them, whether it was about a new baby, a death or ongoing concerns about their environment.

Finally, after so much work, I felt the dolphins had a future. I hoped the sanctuary would ensure their safety and survival.

Once again, however, I had jumped to the wrong conclusion.

19

MORE
LOSS

WITH THE SANCTUARY established in 2005, dolphin mat-
ters briefly took second place. Domestic practicalities took
over.

From 2004 my profile in Adelaide had risen, as had my
mother's involvement in my life. She appeared to enjoy having a well-known daughter and boasted to her neighbours
about how proud she was of me. I wanted to believe it but,
sometimes, it seemed contrived. Over the years, I'd come to
accept that my mother was never going to be who I wanted
or expected her to be. Of course, I loved her; we had a shared
history and we'd been through hell. Even though I was now
in my mid-thirties and maturing as a woman, I still felt
obliged to protect and look after her, as I had when I was
younger. At times she looked like a little girl, lost, just as
she had sitting on the doorstep at Brighton on the evening
I told her we had to leave.

She'd left her partner Sid and was living in a community
housing flat. She met a new man in an adjoining flat and,

eventually, they married. When I became financially secure, I bought her a little brick house with a compact garden a few suburbs away from me. I also bought her a car. She moved into the house with her husband.

I spent my weeks working on environmental stories, hosting charity events and presenting the weather on TV. Most weekends, I prepared meals for my mother and her new husband. Still, I felt nothing I did was ever quite enough. Then, Mum's husband moved out of the house I'd bought them and Mum followed him. I later sold it.

None of these discomforts or distractions meant anything to me, however, compared with what happened in 2007.

Throughout the time I'd been working towards my ten-year career on Channel Seven as the senior environment/science reporter and weather presenter, my brother Mark—the person with whom I'd shared so many of my formative life experiences, from Cornwall to Adelaide—had embarked on a different journey.

He spent time with his wife working overseas in Saudi Arabia. Nurses earned good money there and he mostly enjoyed living in a different culture. But his marriage failed and he returned to Australia with many personal and health problems. He joined an evangelical Christian church in another state and married again. I flew interstate for the ceremony.

As he was getting ready, I was simply his sister and close friend again, who just wanted him to be happy. I helped him with his bow tie. He was wearing an awful faux leather vest.

"Do you like the vest, Mel?" he asked, as I grappled with the rebellious bow tie.

"Yes, Mark. It's very smart." I smiled.

He grabbed my hands mid-fumble. What he said next surprised me.

"Mel, I love you. I want you to know that. We don't endorse what you do for a living. Television and the media are the roots of so much evil in the world. But I know you try to do good for the animals and you look after Mum."

I was taken aback. I wasn't religious. I told myself firmly to be tolerant, especially today, on his big day. I knew—as I'd always known—that his soul, like mine, had constantly been searching for meaning, belonging and love. I hoped, like me, he had finally found it.

He gripped my hands tightly and looked into my eyes.

"You're strong, Mel. You've always been stronger than I am. We've been through a lot. I think I've found some happiness. I know you found happiness with the dolphins. I hope you find it again."

I hugged him. For the last time.

In March 2007, Mark died of a massive heart attack. He was forty-four years old and left behind a young family.

I was overcome with grief. He was the only other person who understood what we'd been through as kids. Although we didn't discuss it often, it was the invisible glue that held us together. It took me a long time to come to terms with losing him. In that time, I reflected on my father, his mental health and behaviour. After much introspection, I knew I finally had to try to forgive him.

The next six years were tough. At the end of 2013, I felt so worn down by work issues, my mother's dramas and the pressure I placed on myself to look after her that I decided to leave Adelaide for a fresh start. My partner Grant and I quit our jobs, boarded a plane, and headed interstate.

Despite knowing how much I'd miss working with Mike and the dolphins, it was a huge relief to leave behind so much stress and so many family ghosts.

Deep in my heart, however, I knew I would reconnect with the Port River one day.

WE EVENTUALLY MOVED to Melbourne, a much larger city than Adelaide. I found casual work in a radio newsroom. I enjoyed learning about this new medium. Compared with TV production, radio is fast and efficient, but the daily grind of turning around short snippets on the hour, every hour, wasn't for me.

I secured a position with the Bureau of Meteorology, going back to something I knew well and that fascinated me—the weather. I relished learning more about the science and enjoyed communicating it to the public, but I disliked the red tape and bureaucracy. Secretly, I pined to be back on the water with the dolphins.

I stayed in contact with Mike and made the occasional trip to Adelaide.

About two years after I left, he published a paper in a scientific journal outlining his research over the previous two decades. It was generally positive, especially about conditions in the river in the Inner Estuary, the area that was declared a dolphin sanctuary.

The paper said that from January 1990 to December 2013 his research team conducted more than seven hundred trips to the river. This averaged thirty-one trips per year where dolphins were counted and recorded.

Between 1990 and 2013 the number of dolphin sightings displayed a "significant increasing trend" and, in fact "the number of dolphin sightings in the estuary doubled during the course of the study."

The paper suggested that improvements in water quality due to a decrease in polluting industries was one contributing factor. The paper also said that enforcement of laws in relation to harassment of dolphins and reductions in boat speeds in the sanctuary had contributed to dolphin population growth.

"We suggest that these changes have yielded a more favourable environment for bottle-nosed dolphins resulting in an estimated 6 per cent increase in sightings, from a near absence of sightings in the 1980s."

The paper was generally optimistic but warned that ongoing protection for the dolphins in urban areas was vital.

But his optimism waned as time went by. Mike hinted to me that he and others were becoming increasingly worried about the effectiveness of the sanctuary. He told me the previous Labor government had disbanded the Dolphin Advisory Board, which oversaw the sanctuary. They'd also slashed the number of rangers.

In early 2021, on another trip to Adelaide, Mike shared some concerning news with me.

"I've crunched some numbers. On average, over the past three years, only one in thirteen calves are surviving. Over the same period, the number of dolphins has dropped by about 30 per cent."

I felt like someone had delivered me a sucker punch.

"Why? What's going on?"

"We don't know why. Could be boat strikes, could be entanglements, could be pollution, or lack of food. We don't know what's killing the babies, and I don't know whether the adults are dying or just leaving."

"Is the media still running dolphin stories?" I asked. "Are you still doing them?"

"Yeah, sometimes."

"Maybe dolphins don't rate well on TV anymore." I heard the sarcasm in my voice about my former profession.

I wondered why, when things were looking so positive a few years prior, the number of dolphins living in the river had declined so rapidly? Why, too, were calf mortality rates so high?

Mike seemed to be at a loss and so was I.

But one thing I did know was that I needed to bring the issue to the public and the politicians' attention. I also needed to see the dolphins again, feel inspired by their presence, feel the river and reconnect with it.

I knew this was going to be essential for the next battle ahead.

20

TIME
TO REVISIT

A FEW MONTHS LATER, in July 2021, the forecast was rain. Instead, sunshine slipped through the clouds, blazing like a bright morning star as Mike backed the boat into the water. It was like only a minute had passed since I'd last stood on that ramp, holding the half-cabin in place as it bobbed gently along the gangway.

I looked around. The surroundings were so familiar, but I noticed something new in the distance. Where the old landfill site used to be, a green mound now rose. It had been covered up and planted with grass. Flocks of gulls no longer hovered, squawking and squealing. Amazingly, the old dump site now looked like it had always been part of the landscape.

I closed my eyes, breathing in the brine. My ears filled with the familiar lullaby of boat masts clanging in the breeze. Even the low rumble of the power station seemed soothing.

"Good job, you haven't lost your touch," said Mike as I steadied the boat.

He was carrying a wetsuit and snorkel.

"Are you really going for a dip?"

He threw the gear into the boat.

"I may have to. It's quite possible we'll find a couple of bodies today. If we do, I'm going to have to jump in and collect them. They often hang around this part of the river."

He'd called me a few days before to tell me that two well-known dolphins were feared dead. They'd recently been entangled in fishing line. He hadn't seen the pair for days.

The news came on the back of the shocking revelations about mortality rates which Mike had revealed to me a few months earlier.

I'd flown to Adelaide the previous day, determined to help Mike if I could. Mostly, however, I needed to see the river again. I needed to experience and understand how it had changed first-hand, which would help me continue lobbying the government. I had also written a story about Jock and the plight of the dolphins for the national *Weekend Australian Magazine*. The overwhelming support from readers prompted me to contact the SA government Environment Minister, directly. I'd been in regular contact since and was pleased he at least seemed concerned and keen to get to the bottom of what was killing the dolphins.

I dreaded the thought of finding a dead dolphin on my first trip out in years, but understood why Mike needed to locate them. We boarded the boat and puttered slowly past Jock's old home. His boat was gone. It had sunk, swallowed by the sea, perhaps in homage to its former constant companion.

I smiled as I remembered him racing up to our boat to greet us.

"Remember that channel?" Mike said, as we passed the entrance to our one-time playground. I blotted a tear from my eye and shot him a watery smile.

"We spent so many hours there, faffing around with Jock. I'll never forget."

We dawdled along in silence, trying to scrutinise the million fingers of mangrove roots jutting from the silt at the water's edge, looking for a still grey form that might be caught up in them. I squinted, cursing the brightness for obscuring my vision. Every floating twig seemed sinister. Bits of flotsam, looking suspiciously like tips of dorsal fins, bobbed in the shallows.

We scoured the area. Nothing. I was both relieved and disturbed.

"Looks like we're not going to have any luck today," he sighed. "Maybe the bodies have sunk. I think we need to go and find you some dolphins to say g'day to."

He pushed the throttle forward and we surged upriver. The water was choppy, the wavelets were ablaze with sunlight, laminating the surface with glinting diamonds, almost too dazzling for my eyes.

Although I'd been here hundreds of times, I noticed things I'd previously ignored. I marvelled at the mottled green Adelaide Hills on the distant horizon. They seemed close enough to touch. I craned my neck to examine the gargantuan hydro towers that materialised from the mangroves. They looked like invading aliens that had been stopped mid-stomp, their advance frozen by some unseen force. The sight reminded me of old sci-fi movies.

Around the bend, dozens of Lego-like houses had popped up, their colours garishly bright and jarring against the water. They had replaced the redundant, decaying wharves. I felt deeply sad. The wharves had witnessed so much history; they'd once played such a vital role in connecting people to the river.

"So many new houses," I said, as Mike carefully navigated the boat through the channel markers.

"Yeah, they're everywhere. I guess it's both good and bad. It's bad 'cause it'll probably put more pressure on the river, and good because there might be more eyes keeping a watch on it."

I understood his logic. We passed the old soda ash factory, which I'd loathed as a student. It was now closed, its white discharge no longer creating caustic clouds in the river. The place looked like it had been bombed. It was a mass of twisted remnants of machinery and charred towers, like something out of Tolkien's Mordor.

"So, the factory has gone."

"Yep, good riddance," he said. "But that's the conundrum. The river's probably cleaner than it's ever been, but the dolphins are dying or moving out. It just doesn't make sense."

As I stared down at the water, I wondered if there were still toxins leaching into the river that were killing the babies, or if they were getting a download from their mother's milk? Were there new toxins entering the waterway from all the urban development I was seeing along the riverbanks? Was it old toxins in the sediment, which was accumulating through the food chain? Or could it be inbreeding causing genetic mutations, being such a small population? Could that have caused defects that made the dolphins' immune systems weaker? Or were the dolphins suffering from a bacteria or disease? I knew more than fifty dolphins in the Gulf St. Vincent had died from a measles-like virus called morbillivirus back in 2013—could it be another outbreak? Research had found dolphins in other parts of the world had also died from similar viruses. Or was it something more conspicuous, like boat strikes; were there more boats on the river? Or were fish stocks so depleted that nature was somehow controlling population growth?

Squatting on the bow, gripping the rail, plagued by these questions, I scanned the water, hoping to see a dorsal fin emerge.

"We may not see any today," I yelled over the noise of the motor.

"I've been out here for thirty years, and I've never not seen a dolphin. Today is not the day I'll break that record. You need to see some dolphins, and we will."

We headed farther out.

I remembered the first time we'd gone looking for dolphins, all those years ago, when three had emerged in front of us before disappearing again. I remembered the circle-shaped "footprint" of still water on the surface when they dived. It had been enthralling to learn so much about these sleek, beautiful creatures. Surely, we'd see dolphins today... wouldn't we?

Suddenly, a thrill shot through me. Ahead, surfacing from a wave's dimple, was the tip of a fin.

"Over there!" I called excitedly.

"Where?"

"Two o'clock. Maybe 10 metres away. I think there's a pod, could be three."

I stood up, feet planted wide, trying to get a better view. "I think one's got a white streak on its fin."

"Right," Mike yelled back. "That sounds like Hope."

How appropriate, I thought. Then, to my delight, the trio of fins changed direction in unison and moved quickly towards the boat.

Mike's head popped up from the cabin. Camera in hand, he started clicking.

"Who's driving the boat?" I said.

"I'm doing it with my foot. It's an old Fijian thing. I do it a lot when I'm out here by myself."

"I can drive if you want to concentrate on the shots."

"Nope. Stay where you are. Look, they're racing over to see you."

He lowered his camera, his eyes glistening. He chuckled. "I think they've missed you."

My heart inflated, filled with an indescribable rapture. The pod congregated around the hull. Another two dolphins appeared from nowhere, joining the group.

"Puh, puh."

God, how I'd missed that sound.

Like synchronised swimmers, they barrel-rolled and spun onto their backs, pectorals splintering the surface. They turned onto their sides, eyes up. I watched them watching me.

Ping. I felt that pure joy and connection hit me. Again.

"Hello!" I called. My smile must have been so broad.

We inched forward. I pushed the hair from my eyes. I wanted to watch every turn, every dazzling fluke flap, every dive. I strained to hear every breath.

"They want a ride," said Mike. "So, let's give them one."

As if on cue, they surrounded the bow, revelling in the swell, turning and weaving and spinning in the froth and bubbles. I wondered, as I'd done many times before, whether it felt like a spa.

Seeing dolphins again brought on a flood of so many memories. I thought back to all those times I went out on Paul's yacht and *Falie* and watched in awe at the dolphins playing beneath me as I lay on the bow. To those days when I'd seen Billie tail-walk and other dolphins play just for the hell of it. And, of course, to my wonderful Jock. The rush of memories brought a smile to my face.

Warmth spread through me. I felt euphoric. I'd finally come home to my family.

THE POD STAYED WITH US for a long time. Eventually, they broke off to pick up a ride on a much bigger wake caused by the bulbous bow of a tugboat. I'd always loved tugboats, so strong and dependable.

We watched the dolphins leaping high into the air from the surging, gurgling, whirling foam. They looked as graceful as any ballet troupe.

These Port River dolphins had changed my life. I'd been friends with them for more than thirty years. Mike, too, had helped me nurture a belief in myself.

I flew back to Melbourne elated—but more concerned than ever.

OVER THE COMING MONTHS, more dolphins became sick and one had to be euthanised by rangers. Over the course of one year, six healthy adult river dolphins became severely emaciated and died. One female, called Ripple, left behind a seven-month-old calf.

Tissue biopsies were taken from some of the dolphins and an interim necropsy report undertaken in late 2021 showed they were suffering from various infections and diseases.

While their cause of death was inconclusive, their immune systems were found to be compromised, making them more susceptible to illnesses. Necropsies on other dolphins found large fishing hooks and line in their digestive systems.

In mid-2022, a Parliamentary enquiry was launched into the issue. I was asked to provide a submission, which I did. Along with Mike and other concerned parties, I presented to the politicians involved with the enquiry. Most submissions, including mine, called for better water and sediment testing in the sanctuary, reduced speed limits, the banning of heavy gauge fishing and live bait and increased fines for industrial polluters.

While more calves have been born, few have survived beyond weaning. It's estimated there are only half the number of resident dolphins in the river that there used to be. However, I remain hopeful that the population will bounce back, as long as there is a strong will to discover what's causing the dolphins to die and to make legislative changes that will eliminate threats to them and their environment.

Few places in the world are lucky enough to have dolphins living so close to a capital city. This proximity hasn't always been a blessing for them but, with effort, it can be.

The Port River is a natural wonder. It's been taken for granted and abused for too many years, yet is so resilient and breathtaking. The community, government and industry need to pull together to ensure the future survival of the Port River dolphins, to secure their protection and preserve what's left of their natural environment.

I hope it's not too late for Adelaide's Port River dolphins. With effort, it could be a special place where humans and animals coexist, respectfully. Where we can experience the natural world, connect, learn and even heal. A true sanctuary.

For my part, I know that words have weight and stories can change the world—and that is one of the reasons why I decided to write this book. I have always been passionate about protecting marine life and helping to ensure the welfare of marine mammals.

As I prepared my memoir for publication in North America and the UK, Dr. Jane Goodall DBE offered me the chance to be involved in the Jane Goodall Institute's cetacean committee. The committee will explore bringing captive dolphins and whales back to a more natural environment, either marine sanctuaries or reintroductions into the wild. I'm excited about the prospect of continuing to help these beautiful mammals—this time internationally.

21

FINAL
REFLECTION

IN THE YEARS SINCE I swam with Jock, attitudes have changed towards human interaction with wild dolphins. We have become more enlightened and realised that close relationships with these remarkable, intelligent mammals aren't always beneficial for them. They sometimes suffer at the hands of human maliciousness and stupidity, their trust in us cruelly broken. But I like to believe that most people love, admire and want to protect the world's marine mammals.

For me, Jock was, and always will be, the most special of all the dolphins. His unconditional acceptance and trust obliterated the titanium plates around my heart. With Jock, I felt as if I had experienced love in its raw, unfiltered form. His loneliness and detachment from his own kind had mirrored mine. I had connected with them. His physical scars had reflected my internal ones. I'd found my soul mate in the sea. I felt he saw me in a way I'd never seen myself. For him, I felt I was someone special, worthy of spending time with—no strings attached. It was a pure experience without preconceived expectations, emotional manipulation or baggage.

Jock, the dolphins and the river's wildlife helped me delve into the deep-seated feelings that I had squirrelled away and hidden. They showed me the beautiful simplicity of living in the moment, accepting what is, forgiving what isn't and acknowledging what will never be. Connecting with the natural world also helped me find a purpose, a path and passion I doubt I'd have discovered otherwise. I was lucky enough to be guided through a special portal into another world—one that encouraged me to think beyond myself, suspend self-doubt and reconnect with something real—far more important and precious. I could focus on something other than my own sad and, at times, shocking past—which had been dominated by violence, fear, sorrow and loneliness.

I realized that I was not defined by my past. I could choose my future. Like nature, my life could evolve. I was the only person responsible for that transformation. I had the power to make choices. In the end, Jock didn't allow his past, his scars or injuries to stop him from trusting others, I learned that I too needed to allow myself to be vulnerable enough to confront and conquer my fears.

Jock's boldness in finally venturing out of his comfort zone to connect with other dolphins taught me that courage is the most important quality in life. I once heard that you're often at your bravest when you face your fears. It's true. I've faced mine more times than I can remember. But I know I'm stronger for it. Despite the trials and trauma, I consider myself to be the victor, not the victim. When I smile now, it's genuine.

The Port River dolphins became the most influential teachers I'd ever had. They will always be an important and special part of my life. I owe them a debt of gratitude that can never be repaid.

I AM NOT an academic, a scientist or a philosopher, I am just a woman who has learned a great deal from connecting with nature. As a twenty-one-year-old at university I didn't have a smartphone, the internet or Facebook. I connected with something authentic and tangible. I didn't have the option of turning to social media to seek relevance, guidance or acceptance. Instead, with the help of the Port River dolphins, I finally felt interconnected with the world around me and embraced by it.

When I think back to my dear friend, my best friend, Jock, and the times he rested his snout in my hand, it seemed as if he was telling me: It's okay. You're worthy. You're special, this is real. Everything is going to be okay. I'm your friend.

Of all the people and animals I have ever known and loved, Jock was the one who taught me the most—how to trust.

ACKNOWLEDGEMENTS

I STARTED WRITING this book many times over, but never made it past the first chapter. I simply didn't know how to continue. I felt I didn't have the skills and that my story wouldn't matter. It wasn't until I recently learned of the staggering plight of the Port River dolphins, with which I'd spent so many years, that I felt determined to finish. I poured my heart into this book with the hope it will draw attention to Port Adelaide's dolphins—how special they are and why they need protection. As the dolphins did for me, I hope this book brings some light into the lives of those young men and women trapped in the dark world of domestic violence.

I owe a great debt of thanks to the people who've supported me along this journey. To my partner, Grant Robb, for encouraging me, supporting me and tolerating my emotional outpourings, bouts of frustration and elation. Although you've known my story for many years, you always listened patiently as I relayed it again, every night while writing, knowing you'd much rather be watching TV with a cold beer. Your empathy towards animals is inspirational and I couldn't have done this without you.

To Alan Atkinson, my guide and confidant. Your patience, unwavering encouragement and counsel have been vital.

While many friends told me I should... you were the one who convinced me I could. Thank you for being so pernickety about punctuation and for pushing me to dig deeper, even when I didn't want to. At times, we bickered like an old married couple, but you have taught me so much. You are a true mate.

To Peter Dunstone, the detective who was my ray of light in the dark days following the attack on my mother. Your compassion, confidence and reassurance were crucial, as is your friendship now. I am so thankful we reconnected after thirty-five years. I now consider you, Christine and your daughters members of my extended family.

To my buddy Stephen Sard, for your unflinching belief in me all these years. Through the tough times when I was unsure of myself, you've remained a loyal friend and confidant. You are an extraordinarily talented editor and cameraman and one of the nicest blokes I know.

Thank you, Andrew Tupper, for constantly hassling me to put pen to paper. You've always said you're a long-term card holder of the "Mel Support Club" and thank goodness you are. Thank you for listening to and believing in me. I especially appreciated your DIY skills when you came over to assemble a flat-pack desk in my spare bedroom. I spent months at that desk during the world's longest COVID lockdown drafting this book and, to your credit, it's as sturdy as ever.

To Mike Bossley, my friend for more than thirty years. Thank you for being my mentor. You introduced me to a world that changed me forever. Without you I would never have met the dolphins, learned to love the river, or appreciated nature as I do. Although I was just one of your many research assistants, you should never underestimate the impact you had on me.

To Angelique Smelt, my long-time friend, for being the voice of positivity on the other end of the phone as I faced empty pages and constant COVID lockdowns. Your emotional support packages, filled with my favourite chocolates from back home, provided moments of sheer bliss. Your messages of love and support touched my heart.

To my other dear friends and valued colleagues in Adelaide and across Australia, I sincerely appreciate your messages of encouragement. Many were profound and beautiful.

To the team at my Australian publishers, Allen & Unwin. Thank you for believing in my story and taking it on. I couldn't have asked for a more genuine and professional group of people to work with. Also, many thanks to the fine folks at Greystone Books for bringing this story to Canada, the UK and the US. I feel honoured to be part of your stable of inspirational, engaging authors.

Finally, but just as importantly, a huge thank you to Q, my big black rescue cat and constant companion. You were always by my side and ready to provide a wet nose nudge when you saw the tears spilling as I wrote. As well as being my furry alarm clock, you never forgot to remind me when lunchtime had arrived, and we both needed a break and snack. Like all the animals in my life, you are a source of pure delight, inspiration and unconditional friendship.

HOW YOU
CAN HELP

WHILE I BELIEVE lobbying government is the most effective
way to push for legislative change to improve outcomes for
marine life, I have listed below some charities you may want
to consider if you'd like to donate or help. I am not affiliated
with any of these organisations.

ANIMAL WELFARE INSTITUTE
general: https://awionline.org/
donate: https://awionline.org/content/make-donation-awi

Animal Welfare Institute is a US-based charity with a
focused campaign to keep dolphins and whales out of
captivity.

HUMANE SOCIETY INTERNATIONAL (AUSTRALIA)
general: https://hsi.org.au/
donate: https://action.hsi.org.au/page/34353/donate/1

Humane Society International works towards a more
humane and sustainable world for animals. It also has a
long-term campaign to keep dolphins and whales out of
captivity.

HUMANE SOCIETY (US)

general: https://humanesociety.org
donate: https://www.humanesociety.org/
   how-you-can-help

The Humane Society of the United States aims to end suffering for all animals and make the ocean safer for all those who call it home.

JANE GOODALL INSTITUTE

general: https://www.thejanegoodallinstitute.com
donate: https://www.janegoodallinstitute.com/contact

The Jane Goodall Institute is a global organisation that inspires people to conserve the natural world we all share, improving the lives of people, animals and the environment.

OCEANA

general: https://oceana.org/
donate: https://act.oceana.org/page/73742/donate/1

Oceana is the largest organisation in the world solely devoted to marine conservation. It has a long-term focus on reducing the impact of fisheries on marine ecosystems.

OCEANCARE

general: https://www.oceancare.org/en/
donate: www.oceancare.org/en/support/donations/

Most of OceanCare's donations are invested directly back into projects for animals in the wild or for campaigns fighting for their protection.

ORRCA
general and membership: https://www.orrca.org.au
donate: https://www.orrca.org.au/join-our-rescue-team

ORRCA is the Organisation for the Rescue and Research of
Cetaceans in Australia. It is a volunteer organisation with
a vision to help rescue, research, conserve and protect
the welfare of marine mammals in Australia. It is always
looking for new members to train as rescuers.

TETHYS RESEARCH INSTITUTE
general: https://tethys.org
donate: https://tethys.org/support-us/

Tethys Research Institute is a non-profit research organization supporting marine conservation through science
and public awareness. Its work has moved the general
public attention towards the Mediterranean cetaceans,
and to better understand how to mitigate the pressures
that threaten these wonderful animals.

WHALE AND DOLPHIN CONSERVATION (WDC)
general: https://au.whales.org
donate: https://adopt-au.whales.org/donation-form/

WDC has a long-standing program to encourage the
recognition of the rights of whales and dolphins within
global and local conservation policy. The Australian arm
supports Dr. Mike Bossley's work.

MELODY HORRILL is an award-winning environment and science journalist, presenter and documentary producer. She is well known for her passionate writing about and filming of wild dolphins in Australia. Her documentary *A Dance With a Dolphin* was broadcast around the world on CNN. Horrill is a member of the Jane Goodall Institute's cetacean committee and a director of the Dolphin Research Institute in Australia. *The Dolphin Who Saved* Me is her first book.